ECONOMIC IMPACT OF LARGE PUBLIC PROGRAMS:
THE NASA EXPERIENCE

ELI GINZBERG
JAMES W. KUHN
JEROME SCHNEE
BORIS YAVITZ

Olympus Publishing Company Salt Lake City

338.475
N 241

Library of Congress Cataloging in Publication Data

Main entry under title:

The NASA story.

Bibliography: p.

Includes index.

1. United States. National Aeronautics and Space Administration. 2. United States -- Appropriations and expenditures. 3. Aerospace industries -- United States. I. Ginzberg, Eli, 1911-

TL521.312.N37   338.4'7'50090973   76-5467
ISBN 0-913420-68-9

Contents

LIST OF TABLES

LIST OF FIGURES

Acknowledgments

A collaboration between the Conservation of Human Resources, Columbia University, and NASA requires some explanation, for the Conservation project is best known as a research group with specialized skills in the area of human resources and manpower.

The initial contact between NASA and the Conservation project was made by Dr. Robert Jastrow who, in addition to serving as the director of the Goddard Institute for Space Studies, is also a member of the faculty of Columbia University. In 1971 Dr. Jastrow first discussed with me a suggestion he had made a few days earlier to his NASA colleagues that the Conservation project might help in one of their new undertakings — an exploration of whether it was possible to develop a realistic assessment of the economic impact of the entire NASA program. Jastrow and I had been members of a Columbia seminar on technology and social change some years earlier, and I had edited the seminar papers and discussion for publication (1964) under that title. Even more relevant was the Conservation project's earlier work, *The Pluralistic Economy*, which broke new

5

ground in explaining how large-scale public expenditures influenced the shape and direction of the national economy.

Other members of the staff had periodically dealt with scientific or technological matters. Specifically, James W. Kuhn had written an account of the role of technical personnel in the diffusion of nuclear energy, and had also done some consulting for NASA. Boris Yavitz had authored two volumes on computers, the most recent with Thomas Stanback, Jr., *Electronic Data Processing in New York City*. Jerome Schnee, who joined us specifically to work on the NASA project, had been most directly involved in the study of technology and economics, notably as a co-author with Edwin Mansfield and others of *Research and Innovation in the Modern Corporation*.

These five — Ginzberg, Hiestand, Kuhn, Schnee, and Yavitz — became the Conservation project's NASA team. Schnee was responsible for the largest part of the active research and for preparing most of the final manuscript. However, the group worked as a team in the difficult task of designing the several approaches which were pursued during each stage of the project. It is no disparagement of the contributions that the others made to say that Boris Yavitz provided leadership in the complex design problems.

An early discussion between Dr. George Low, deputy administrator of NASA, and me confirmed that the senior NASA staff were deeply interested in having a valid impact study conducted and that they would play an active role in helping the research team. Once Dr. Low's interest and participation were confirmed, the Conservation group was eager to undertake the assignment. We recognized that active participation by the deputy administrator, who had played a key role in the Apollo project, could not fail to provide significant lessons. And that is exactly how things turned out. Throughout the three-year study, the Conservation group met approximately every three months with Dr. Low (with various other senior NASA staff members also attending). Each session lasted several hours and consisted of reporting by the Conservation staff on its activities, reviewing with Dr. Low the issues about which Conservation staff needed enlightenment, and exploring the most sensible next steps. The best way to describe what happened during this three-year period is to emphasize that the Conservation group had a continuing seminar with Dr. Low. Without his active participation this research effort would have faltered. There were too many places along the road where a group of outsiders, acting alone, would have lost their way.

In the preparation of this study for publication, two members of the Conservation staff made substantial contributions. Dr. Charles Brecher worked with the authors on revisions of the final manuscript, and Ms. Anna Dutka carefully reviewed and corrected the page proofs.

Eli Ginzberg
March 1976

Foreword

A foreword is often decorative, little more. The reader can skip it without loss and get right to the heart of the study and findings. But this foreword has an essential function: These introductory pages provide a framework for the case study of NASA. Although NASA is interesting in its own right, it is the framework that extends its relevance and adds significance to the findings.

The conventional view is that the dynamism of the American economy resides in the private sector. The investments that the private sector makes, the employment it generates, the income it distributes, and the taxes it pays underpin the growth of the economy and the well-being of the American people. Clearly, there is much to support this view. Yet it is exaggerated.

There are large public programs in such diversified fields as highways, agriculture, education, research, medicine, space, and manpower that perform a series of economic functions not sharply distinguishable from those of the private sector. Moreover, these large public programs play an ever larger role in giving shape and direction to the American economy and society. In 1929 the not-for-profit sector accounted for about 10 percent of the national income; by 1975 the comparable proportion had reached about one-third.

The importance of the public sector was the theme of an earlier study by members of the Conservation staff, *The Pluralistic Economy*. In that book my collaborators and I demonstrated the scale and significance of not-for-profit activities and argued that a realistic framework for analyzing the American economy required attention to the principle of complementarity. The private and public sectors had become increasingly intertwined and could no longer be treated as separate entities, each operating under its own logic, responding to different stimuli, and yielding distinctive products. While profit, nonprofit, and government institutions differ in their entrepreneurial roles, each is often engaged in producing useful goods and services, is constrained by revenues, and must obtain labor and supplies in the market.

Considering the importance of these large public programs which in total account for an estimated two out of every five jobs, we find that they have commanded very little attention. American economists have limited their concerns to dollar flows through the system, ignoring for the most part its institutional structures.

This book represents an elaboration and refinement of the earlier study. The focus is on a single case — NASA — wherein approximately $40 billion were spent within one decade with the specific mission of getting a man to the moon. But during a similar ten-year period, comparable or larger public expenditures can be identified with a wide range of missions.

The question to be confronted is: What generalizations can be presented for the study of public programs that range from Medicare to space? In short, what kind of framework can be created to facilitate the systematic analysis of large public programs? The paucity of earlier research into this crucially important new dimension of the American economy precludes this foreword from more than sketching the outlines. It will require many investigators, working over a long period of time, to develop a model of public program expenditures that will even approximate that used for illuminating private sector activities.

The following seven points represent a preliminary check list of the relevant dimensions of a large public program, which may become the building blocks for an analytic framework. First, has the relevant unit of government (federal, state, or local) been previously engaged in the activity, or is it substantially a new undertaking? For example, new defense programs can be viewed as extensions of current functions, but others are entirely new functions for particular levels of government, such as federal manpower training after 1962, or the assumption of financial responsibility for the medical care of older persons with the passage of Medicare in 1965.

Second, what was the source of initiative for the launching of the program? There may have been a number of strong interest groups that had long lobbied for the new programs, as with Medicare; or it may have been largely a Presidential initiative, as in the case of NASA.

Third, to what extent is the new program delineated in the sense that it has a specific goal? The Works Progress Administration of the 1930s specifically sought to provide employment and income for the millions of unemployed; in contrast, Medicare represents an ongoing commitment.

Fourth, closely related to the foregoing is the political potential of impact groups, those who are affected by the new or expanded program. The agricultural support programs which date back to the late 1920s and early '30s have been hard to modify, much less eliminate, because of the large numbers of organized voters who might be adversely affected. Medicare beneficiaries also represent a large block of concerned voters who will look askance at any radical effort to reduce the benefits provided by the original legislation. In contrast, it has been possible, not once but several times in recent history, for Congress to reduce radically the budget of the Department of Defense, with the consequent closure of production facilities and elimination of jobs. The result was the abandonment of what were formerly thriving communities.

The most important observations about impact groups with respect to large public programs are: (a) their presence, (b) the essential support they provide in the ceaseless struggle for resources, and (c) part of the price paid for their support is the restricted freedom of the Executive branch and the legislature whose actions must take into consideration the expectations and values of the impact groups.

Fifth, the nature of the infrastructure that must be used in order to accomplish the new program is still another critical element. In the case of federal manpower training programs, there was no feasible way for the federal government to avoid involving two principal institutions — the vocational education system and the federal-state employment service — a constraint that seriously influenced what the new effort could and could not accomplish. In the case of Medicare, physicians, hospitals, commercial and nonprofit insurance, and pharmaceutical companies were hardly passive bystanders. They played a major role in shaping the legislation and its implementation.

Sixth, every large public program, no matter how innovative it initially appears, tarnishes with time if only because the resources that it consumes, if released, could provide the wherewithall to launch an alternative program. Hence every large public program becomes increasingly vulnerable with time and probably will not be able to continue growing at its early rate. A major challenge to the concerned administrators, the impact groups, and the involved legislators is to explore continuously how the original program goals can be transformed so as to permit the government agency to help assure the continued well-being of the various impact groups. The experience of three large public programs of the 1960s — NASA, Medicare, and manpower — are indicative of this process of goal transformation.

Although NASA had from the outset the specific mission of putting a man on the moon, before the Russians could get there (it was hoped), the project was clearly perceived, at least by its first administrator and his closest advisers, as also having the more open-ended mission of space exploration. Similarly, the Manpower Development and Training Act (MDTA) program was only one year old when Congress expanded its mission from the retraining of skilled workers, who had presumptively lost their jobs from automation, to the much broader goal of assisting all those who were encountering difficulties in the world of work. To reinforce this shift, Congress waived the requirement of state contributions and substantially enlarged appropriations to give increasing evidence of its long-term commitment to manpower development. The circumstances were not so different in the case of Medicare. Many of the proponents, in and out of Congress, looked upon its passage as simply a first step toward a comprehensive system of national health insurance.

Seventh, within the federal arena, important networks of political alignments are crucial to the launching and survival of large public programs. For many years federal expenditures for biomedical research increased rapidly as a result of the intimate alliance among the leadership of the National Institutes of Health, Senator Hill and Representative Fogarty, the respective chairmen of the appropriations subcommittees of the Senate and the House, and the principal beneficiaries of the research grants who provided the expert testimony which the chairmen used to persuade their colleagues that the enlarged appropriations were being well used. This linkage among impact groups, congressional leaders, and the responsible bureaucracy provides the dynamism for ongoing support for many large programs, even after the interest of the President and public may have waned. (It might be noted that the brutal dismissal of so many senior officials at the end of the first Nixon Administration was predicated on the theory that the President would be in a better position to be the sole ruler of the bureaucracy if the preexisting links between his political appointees and their own staffs and between the departments and the congressional leaders could be severed.)

Folklore has it that the bureaucracy is responsible for the survival of many programs that are no longer needed. While the bureaucracy unquestionably often plays a role, the true answer must be sought in the powerful alliance between impact groups and key legislators.

In summary, the building blocks that we have identified for an analytic framework for the study of large public programs involve the following:

(1) The nature of the program, more particularly whether it represents a new governmental activity or a change in scale of an existing function.

(2) The source of the initiative: Is it located in the Executive branch, the Congress, or among special interest groups?

(3) The goal of the program: Is it specific and time limited, or general and ongoing?

(4) The characteristics of the principal impact groups; that is, the identifiable beneficiaries of the new program, both producers and consumers.

(5) The extent to which existing infrastructure can or must be used to implement the new program; or whether and to what extent an infrastructure must be created.

(6) The potential for transforming the goals of the program once broad-based support has begun to wane.

(7) The strength of political alignments among the bureaucracy, congressional leadership, and the impact groups.

To what extent can these building blocks be used to illuminate selective aspects of large public programs? A first effort follows by focusing on two elements in the schema — the transformation of goals and the role of the existing infrastructure — for three programs launched in the 1960s — manpower, Medicare, and NASA.

As earlier suggested, the effectiveness of the federal manpower efforts during their first decade was substantially constrained by the capabilities and limitations of two principal structures — vocational education and the federal-state employment services. It was impossible for the federal government to launch a large-scale training program for the disadvantaged except by developing a working alliance with vocational education, the only system that had nationwide scope and experience in skills instruction within the classroom.

The fact that the leaders of the field had built a system that generally was deficient in providing a variety of training was beside the point. There was no other infrastructure available, at least initially, to meet the new program's needs. And it should be noted that vocational education put aside its aversion to dealing with the disadvantaged and responded with considerable speed, enthusiasm, and good will to the new program. The leadership recognized that the substantial new flow of federal funds offered it an opportunity that it dare not neglect if it wanted to remain in control of its field. True, the leadership made clear that it preferred the new funds to come directly to vocational education rather than through manpower channels; but when they found that Congress was not interested in heeding this advice (for fear that the poor would again be dealt out), the leaders went along with the new program, with only occasional sniping from the sidelines.

The employment service — with its multiple functions of registering and paying benefits to the unemployed, providing testing and counseling for new and hard-to-place applicants, and serving as a placement agency for the unemployed and those who sought new and better jobs — had long been the bane of many federal officials who were unable to control it. Moreover, the employment service had a generally poor reputation among employers, organized labor, and, in particular, minorities. Accordingly, in the early days of MDTA, senior Labor Department officials explored whether the new program could be launched without relying on the employment service for assistance. These efforts were largely stillborn because only the employment service had a network of suitable offices throughout the United States. Whatever the shortcomings of the employment service — and there were many, although a large proportion of the criticism was unrealistic — the leadership of the Labor Department found itself without an effective alternative. Labor Department officials did the only thing possible and worked hard to modify the focus and operations of the employment service so that it would be more responsive to the disadvantaged and would play a leading role in the referral of eligible persons to the new training programs. There are many who believe that the Labor Department succeeded to such an extent that the employment service, in its new effort to serve the disadvantaged, lost its capability to meet the placement needs of its regular clientele, and thereby put itself in an untenable long-run position.

The crucial point of this review of manpower infrastructure is to emphasize on the one hand that the new national program had to rely on the two principal institutions in place at the time when it was launched, and on the other hand that over the following years the federal government was able by strenuous efforts to modify the operations of both, even while putting new structures into place.

The Medicare story perhaps reveals even more about the role that the existing infrastructure plays in the shaping of new public programs. The dominant physician lobby had successfully resisted any large-scale government insurance system until 1965 when Congress, with the President in the lead, could no longer ignore the large number of middle-class and working-class voters who saw the risk of bankruptcy hang over their heads should their parents in their old age be stricken by a serious illness. But even in its moment of defeat the physician lobby had considerable power: Congress agreed that when treating Medicare patients, physicians would receive their customary fees and would be under no compulsion to alter in any way their conventional pattern of practice.

The hospitals also made their influence felt. They convinced Congress not only to allow full reimbursement, including depreciation, but to provide an additional 2 percent override, ostensibly to help hospitals modernize and keep abreast of new developments.

Nor was the insurance industry without influence. It succeeded in winning for itself a preferred role as the fiscal intermediary between the Social Security Administration that paid the bills and the hospitals and physicians who provided the services. Moreover, several private insurance companies developed policies specially designed to supplement Medicare coverage.

In sum, physicians, hospitals, and insurance companies were all well taken care of under the new legislation. That is, initially. It required only two years before the Social Security Administration recognized that its initial cost estimates were seriously awry and that if future costs were to be kept within bounds the law would have to be amended. The next years saw the first steps at cost control. The hospitals' 2 percent override was the first to go. Then limits were placed on fee reimbursement. Later, the Social Security Administration sought to prevent the construction of an excessive number of beds by limiting its reimbursement to those new beds that had been approved by the local hospital planning agency. Eventually, steps were taken to encourage physicians to prescribe generically rather than by brand name. And the federal government devised additional rules and regulations all aimed at keeping its financial commitment under control.

To the extent that its goal was directed to relieving old people of financial burdens growing out of high medical bills. Medicare has had considerable, but not phenomenal, success. In 1973 Medicare covered approximately 40 percent of the total medical expenses of the eligible age group, but this is a smaller share than in its first years. This reduction in relative coverage has been accompanied by repeated increases in premiums paid by the participants and much larger drains on general revenues than the actuaries had anticipated.

A more interesting aspect of Medicare in the present discussion of infrastructure is the steady compulsion that the federal government has been under to alter the shape and operations of that infrastructure because of the financial burden which it assumed. Reference has already been made to the government's concern with bed control, prescription practices, and fee structure. One should also note its insistence on rationing new expensive services such as dyalisis. Further, the terms under which it will reimburse hospitals for patient care have led to the elimination of large wards in teaching hospitals.

The government is currently locked in a major dispute with medical schools and their teaching hospitals over the conditions under which it will pay for physician services. It is possible that in the near future, the federal government will use its reimbursement practices to alter the number and distribution of residents in training for the several specialties. While the consequences cannot be foreseen at this early date, the passage in 1972 of the Professional Services Review Organization amendment to the Social Security Act may have launched a national experiment in quality assurance in the delivery of medical care.

In short, what started as a social insurance scheme to cover the high costs of illness among older persons is today, as the American Medical Association foresaw, transforming the delivery of health care. The issue in the present perspective is not whether such a transformation will turn out to be a gain or a loss but rather the extent to which quite specific and delimited goals are transformed in the process of implementation and eventually effect changes in the critically important infrastructure.

This brings us to the second element of large-scale public programs where the NASA story suggests generalizations: the transformation of goals. No public program, no matter how strong its initial support and how liberal the funding it is able to secure, can exist for more than a short time before some major unit of government or some important group in the community — scientists, journalists, specialists — will raise questions about the relevance of the undertaking or the efficiency of its operations. One significant parallel among space, manpower, and Medicare is the relatively long period of friendly toleration that each received from both legislators and the public.

But sooner or later the period of grace is certain to come to an end as the benefits, costs, and secondary consequences of the new program come under more careful scrutiny. There is no possibility of escape because Congress is committed to annual appropriations, which means that it will be under pressure to make a periodic reevaluation. In the case of NASA, this became inevitable once Apollo had achieved its mission of landing a man on the moon. Likewise, MDTA escaped critical scrutiny until the reelection campaign of 1972, when most Great Society programs came under attack by conservative Republicans who argued that one cannot solve difficult problems by throwing money at them. But not even the most conservative politicians contemplated the abolition of Medicare. Rather, the battle lines were formed about the rate at which the beneficiaries should contribute through co-insurance and the rate at which new and costly benefits should be introduced.

The more time that has passed since the launching of a new program, the more scope there is for critics. There has been a repeated challenge to NASA to explain how the space shuttle is to be used and the results that can be anticipated. Many Congressmen believe that $3 billion a year for space explorations, with no specific objectives in view, is much too high. But they are still in the minority.

The criticisms of manpower training cut deep. The argument was made that the roughly $3 billion spent annually to train the disadvantaged had an imperceptible effect on the national unemployment rate and that it was unclear whether those who were trained did conspicuously better in the labor market than those who were not. Criticisms of a somewhat different order related to the inability of the federal government to be constructively involved in designing and overseeing more than ten thousand specific projects every year. Congress ignored the first argument and acted on the latter. In 1973 it passed the Comprehensive Employment and Training Act (CETA) which allocated federal funds to cities and states that henceforth would have the primary responsibility for designing and running the program. The several amendments passed by Congress in 1974 increased the federal government's capacity to respond to rising unemployment and underscores Congress's belief that the nation required a strong manpower policy and could not return to the pre-1962 situation.

In the Medicare arena Congress has sought to respond to the accumulating evidence of excessive charges, poor quality, and wasteful duplication by seeking ways to control these excesses while considering its next step in the structuring of a program of national health insurance. As of 1976 the congressional leaders recognize that there is no easy answer and that whatever they do, including biding time, will be the prelude to further difficulties. The same understanding is widespread among well-informed persons in the medical community. For a long time to come the American people will be forced to struggle with the prevailing system, if only because of the wide gap that remains between its aspirations and the present reality.

Large public programs require a sizable commitment of resources. At some point in time it may be relatively easy to gain the consensus required to secure the resources. But in the scramble that always exists among different interests for a larger share of the public revenue, it is inevitable that the more resources a public program commands, the greater its vulnerability.

These general considerations about large public programs provide a backdrop to this book, which focuses on the experience of NASA, perhaps the most dramatic new program in the post-World War II experience of the United States. The remainder of this foreword deals with the delineation of this study and the manner in which the study evolved. Further, it provides a preview of the chapters that follow, including a few of the important lessons that can be extracted from this case study that bear on the improved management of large public programs.

In 1971, the senior officials of NASA began to ask themselves whether it was possible to develop a realistic assessment of the economic impact of the entire NASA program. They sought a valid methodology and hard data that would enable NASA to inform interested outsiders — Congress, the press, the public — of the economic impact of the more than $40 billion that represented NASA's total outlays in putting a man on the moon. After this spectacular goal had been achieved, questions arose with increasing frequency about the "value" of the space effort, particularly because many other social objectives — from eradicating poverty to rebuilding the cities — were lagging as a result of a serious shortfall in federal funding. The initial intent of NASA officials was to learn more about how the agency's large expenditures had affected the American economy and to make this information broadly available.

The hidden assumption that underlay the interest of the NASA officials in a new impact study was their belief that such a study would reveal a very large multiplier — second-order effects that disproved the simple thesis that all government dollars spent had comparable effects. They had developed this strong conviction as a result of their own direct experience, supplemented by the findings of earlier NASA-commissioned studies that pointed to diverse impacts, including the development of new commercial products, the diffusion of new technology and management techniques into the private sector, the training of specialized personnel, the expansion of aerospace infrastructure, and other contributions.

The Conservation project undertook to explore the NASA officials' thesis without making any a priori judgment about whether it would be substantiated. After several months of seeking to develop an appropriate methodology to assess NASA's economic impact in "macro" terms, the Conservation staff concluded that this was not the most productive way for the group to proceed. It reached this conclusion on the basis of a critical review of the prior impact studies. This review of prior efforts is presented in chapter 1.

The implications of the previous efforts after review became clear. While NASA might eventually have easily recognizable second-order impacts of major proportions reflected in the growth of an important new industry, the limited time dimension (one decade) precluded the likelihood of firmly establishing such a development. Moreover, the Conservation group questioned whether the interaction among new technology, new products, and new markets was such that one could ever ascribe to one particular effort such as NASA the sole or predominant credit for the development of a new industry. The more the Conservation group considered the developmental processes involved in the growth of such major industries as railroads or chemical, electrical, automotive, and aeronautical products, the more it doubted the feasibility of establishing a simple one-to-one relationship between a specific historical occurrence and the emergence of a new industry. There were too many intervening factors at work, mutually affecting one another, to enable the student, even with the advantage of historical distance, to relate a major outcome to *one* specific event.

At this point the Conservation group suggested a different methodology — one that would focus more on understanding the dynamics of impact than on the development of hard economic data. The alternative methodology was a case study approach, aimed at probing more deeply into the different meanings of impacts in selected areas where informed persons had alerted us that NASA had left its mark. The three areas chosen for analysis were: the computer industry, astronomy, and weather forecasting.

The single largest commitment of the research staff during the three years of the project was directed to working up the three major case studies. Aside from the usual difficulties that researchers encounter — the absence of significant quantitative data, the inability to reconstruct fully the crucial forces operating at particular decision points, the hesitancy of various interest groups to reveal information that they believe may put them at risk — we ran into one special problem that warrants attention and reflection. The simplest way of describing the problem is this: All persons well informed about the agency's impact are likely to have played a decision-making role at an earlier stage. Moreover, each did so from a special vantage: as a government administrator, as a member of the industry, or as an academic with close ties to one or another of the organizations involved. Hence there is no informed neutral; he does not exist. Each expert who recounts what transpired does so from a restricted viewpoint and from a specific value position.

It took the Conservation group some time to become fully aware of the subjectivity of the evidence that was accumulated in studying the impact of NASA in these three areas; and it took even longer to determine where the predominance of the evidence pointed. The reader should be alerted both to the difficulties that were encountered in collecting the evidence and to the inevitable errors which probably remain, despite the conscientious effort of the Conservation staff to sift and weigh it carefully.

Chapter 2 is the first of the three case studies. It is devoted specifically to the influence of NASA on the growth of semiconductors and the computer industry. The chapter emphasizes at the outset that it is necessary to treat the federal government as a unit, and that it is the combined influence of the Defense Department and NASA that is being assessed, with the former accounting for most of the expenditures. The analysis is centered round three economic aspects: (1) the role of the federal government in financing research and development in this area, (2) the importance of the government as a market, particularly in the early stages of an industry's growth, and (3) the extent to which government policy encouraged the entrance of new firms and permitted the survival of companies that might otherwise have failed. The chapter also deals with the contribution of the Defense Department and NASA to the acceleration of technological changes and the development of specialized manpower on which the progress of the industry largely depended. The chapter concludes that the federal government played a major part in enabling the American computer industry initially to assume the lead in this field, a position that it has maintained since.

Chapter 3 has a different focus. Its probe is concerned with the influence of NASA on what can be called the "rebirth" of the oldest of the sciences — astronomy. The last two decades have been a period of unparalleled discovery in this field. The chapter emphasizes that the breakthrough in radio astronomy predates NASA, but it was both NASA's needs and NASA's liberal financing that led to the transformation of the small field of optical astronomy to a large-scale science based on satellites and other advanced technology in which the numbers being trained and the resources at their disposal grew by several orders of magnitude.

The story of this transformation is told in considerable detail in a constant search to uncover the role that NASA played, but with great care not to equate the large resources that it made available to the significant results achieved. There can be little question that NASA was responsible in many different ways — from providing the resources for the training and employment of new doctorates to stimulating the renewal of geodosy and planetary studies — for the rebirth of astronomy, a field of human inquiry that had for a long time demonstrated relatively little initiative in theory or discovery. Scientific leadership opposed the large budgetary support for astronomy from NASA out of fear that too many dollars would be redirected away from high-priority areas in physics and biology. But some of the scientific opposition was grounded in the belief — later proved to be grossly mistaken — that little new scientific knowledge would result from space exploration.

The story of the renaissance of one of the oldest branches of human inquiry is exciting on its own terms. But in the context of this study it presents overwhelming evidence that within a single decade the federal effort in space had permanently altered astronomy — i.e., the questions astronomers asked, the available data with which they could work, and the tools with which scientists carried out their work. While the more important results from this rapidly accumulating body of new knowledge remain to be revealed in the near and distant future, the broadening and deepening of understanding about the universe are without recent parallel.

Chapter 4 presents the results of the third probe, an analysis of the economic impacts of meteorological satellites. The chapter starts with a review of the recent advances in meteorology, beginning with the launching of the first weather satellite, Tiros I, in 1960. It then considers the extent to which limited computer capacity restricted the processing of the large amounts of new data that became available as a result of improved observations and transmissions. Next, the focus shifts to explore the evidence of the economic value of improved weather forecasts and points out that there is a considerable underutilization of currently available forecasts and that certain highly distinctive weather phenomena, particularly tornadoes, cannot as yet be predicted with a lead time sufficient to be useful in the protection of property. The author concludes his chapter as follows: Meteorological satellites have transformed weather observation but have not as yet led to improved accuracy of weather forecasting, although it is reasonable to expect gains to be made in the future.

Before the Conservation group's work was finished, the members made still one more shift in focus. While the three probes had provided considerable understanding of both the potentialities and limitations of impact studies, and of NASA's influence in the development of the computer industry, astronomy, and weather forecasting, a general methodology had not been devised that could guide future efforts in this area. Consequently the Conservation group made a strenuous effort to develop an analytical framework for the study of large public programs and, further, to illustrate how such a framework could prove useful to the administrator of such a program. In short, the members sought to make an initial contribution to both the analysis and management of public programs.

The two concluding chapters are directed to extracting some of the general findings contained in the preceding three probes and pointing out how impact analysis can be used by the administrator as a planning and evaluation tool. The analytical framework delineated in chapter 5 provides some guidance to future investigators who seek to bring order out of the complex forces that are constantly interacting whenever large public programs are initiated and implemented. Chapter 6 is an effort to apply this structure to the purposes of management control within NASA. A few highlights from these chapters follow, but no effort is made to do more than suggest the ground that they cover.

The authors remind their readers that every large public program does not start afresh, even one as unique as putting a man on the moon. Clearly, the federal government had to look to the existing aerospace industry to undertake the design and manufacture of the required equipment, to the universities to educate the additional specialized manpower required to staff the many new positions, and to existing as well as new communities to expand rapidly so that new space missions could be carried out. The study of large public programs must therefore begin with an ability to identify and assess the principal institutions that are likely to be involved in the new program and the manner in which they are likely to respond to the setting of priority objectives and the flow of resources to meet them.

A second finding suggests that the nature of the impact of the new program on established institutions will be determined in considerable measure by the relative importance of the new resources made available in comparison to those available from existing sources. Hence NASA had a much greater impact on Huntsville, Alabama, and on the communities around Cape Canaveral in Florida than on Houston, Texas. For the same reason, although it represented but a small part of NASA's total budget, the monies it made available to university departments of astronomy loomed so large in comparison to their regular budgets that NASA's impact was significant in shifting the scale and focus of their operations.

Once NASA was well advanced in implementing its several programs, it found itself with a set of impact groups whose well-being and future were closely linked to its own. While such groups provided strength and support for NASA in the ongoing competition for resources, they also reduced the agency's ability to change its plans or operational decisions without taking their interests into account. Chapter 6 presents a novel and suggestive approach to impact group analysis, drawing on the experience of certain large private companies that have sought to build such an approach into their planning. By sketching the broad outlines of impact group analysis, this chapter provides the top NASA administrators with a tool that, as they shape and hone it, may turn out to be useful in selecting new directions for NASA and in reducing the risks of pursuing them.

This, then, is a preview of what follows. Clearly, the chapters do not make an integrated whole. They are, rather, the products of a research project that found it essential to respond to the consequences of its own efforts and to change direction as it proceeded. But what this book lacks in neatness and symmetry it compensates for in the scope of the questions explored and in the suggestiveness of its answers.

As the first half of this foreword has sought to make clear, public policy is in need of major advances in the analysis of large public programs. This book can be viewed as a pioneering effort directed toward this goal. That the goal is important can be suggested by scanning just three figures — the $40-billion federal expenditures for NASA between 1963 and 1973; the $53 billion for Medicare during 1965 to 1973; and the $16 billion for the MDTA and related manpower programs between 1962 and 1973, making a grand total of $109 billion for these three new programs within one decade!

For better or worse the American economy is today — and will be tomorrow — an admixture of a market sector, a government sector, and a nonprofit sector. It behooves economists, politicians, and government administrators to recognize this fact and to learn as quickly as possible how to improve the planning and management of large public programs so as to better cope with their impacts.

Chapter One

●

Assessment of the
Impacts of the Space Program

As the activities and expenditures of the federal government have grown during the past two decades, the impacts of large public programs have come under closer scrutiny (Solow, 1961-62; Benoit and Boulding, 1963; Oliver, 1971; and U.S. Department of Labor, 1973). Impacts are distinct from the specific accomplishments of an agency in carrying out its overall mission. The placing of a man on the moon or the orbiting of an earth resources satellite are examples of a specific accomplishment. Whereas specific accomplishments relate directly to agency objectives, impacts do not. In achieving a specific goal, a public agency may stimulate the development of a new technology or a new industry, dramatically alter the infrastructure of a city or a region of the country, or create new occupational categories. These ancillary effects are all examples of a program's impact.

The legislation which created the National Aeronautics and Space Administration (NASA) explicitly recognized that the agency's programs would influence the economic and social structure of the nation. The National Aeronautics and Space Act of 1958 directed NASA to conduct long-range studies of the potential benefits from the use of aeronautical and space activities for peaceful purposes. Since its inception, NASA has carried out this mandate by supporting various impact studies of the space program.

27

The studies supported by NASA have examined several types of program impacts. (A preliminary classification of impacts is given by Bauer, 1969, p. 18, and will be reviewed later in the chapter.) These types of program impacts are:

(1) *Economic:* A program may alter the sales and price levels in an industry or the level of employment in a local community.

(2) *Technological:* A program may change the matrix of products, processes, techniques, and materials in an industry or technological field.

(3) *Scientific:* A program may increase understanding of basic phenomena in a particular discipline.

(4) *Managerial:* A program may influence managerial practices and techniques in organizations other than the administrative agency (business firms, other government agencies, and so forth).

(5) *Social:* A program may influence human values and the quality of life.

Previous impact studies have typically dealt with only one of the above. Yet the ultimate significance of a program is that its several different impacts may act on the same institution and, in the aggregate, result in momentous institutional transformations. Because these institutional transformations are of great import, any effort to assess the impact of a large public program such as NASA should focus on the ways in which large public programs modify institutional structures.

The previous impact studies of NASA have never been organized along institutional lines. Thus a first step in assessing NASA's institutional impacts is to extract from earlier studies the significant findings relating to the program's impact on major institutions. This chapter is devoted to a reconsideration of previous studies from the perspective of (1) the total national economy, (2) the local communities that housed major NASA installations, and (3) the industries which sought to meet NASA's technological needs.

NASA's Influence on the Total Economy

The macroeconomic effects of the space program have been considered in four major studies. First, a Midwest Research Institute study (November 1971) used an econometric approach to identify and measure that portion of economic growth — as measured by changes in gross national product — which could be attributed to technological progress. An examination of the relationship between research and development expenditures and technology-induced increases in gross national product indicated that on the average each dollar spent on research and development returns slightly more than $7.00 in gross national product over the eighteen-year period following the expenditure. Assuming that NASA's research and development expenditures have the same payoff as the average in the economy, Midwest Research Institute concluded that $25 billion (in 1958 dollars) spent on civilian space research and development during the 1959-69 period returned $52 billion through 1970 and will continue to produce payoff through 1987, at which time the total gain will have been $181 billion.

A second econometric investigation of the relationship between NASA expenditures and the U.S. economy has recently been carried out by Chase Econometric Associates, Inc. (March 18, 1975, and April 1975). The first phase of the Chase study used the 185 interindustry input-output model developed at the University of Maryland to analyze the short-run economic impact of NASA research and development expenditures. Simulations of the input-output model were conducted, assuming that $1 billion of federal expenditure is transferred (proportionately) from other nondefense programs to NASA with no change in the size of the federal budget. Chase estimated that the $1 billion transfer would increase manufacturing output in 1975 by 0.1 percent, or by $153 million (measured in 1971 dollars). It was also estimated that this $1 billion transfer would increase 1975 manufacturing employment by twenty thousand.

The second phase of the Chase study considered the long-run effects of NASA research and development expenditures. Using a production function which related NASA research and development expenditures to the rate of productivity growth in the U.S. economy over the 1960-74 period, Chase concluded that the social rate of return on NASA research and development expenditures was 43 percent (Midwest Research Institute's estimated social rate of return was 33 percent). This second phase also estimated the effects that changes in NASA research and development expenditures would have on economic growth and stability. Over all, these long-term estimates confirmed the significant positive effects of NASA research and development expenditures on national productivity and employment levels.

A third study of the macroeconomic impact of NASA research and development programs has been conducted by the Space Division of Rockwell International (December 1974). Rockwell investigated the relationship between NASA's space shuttle program and additional employment in the state of California. Using an econometric model developed at the University of California (at Los Angeles), the Rockwell people estimated that the space shuttle program had an employment multiplier of 2.8; i.e., direct shuttle employment of 95,300 man-years in California produced an increase of 266,000 man-years in total employment.

Each of the econometric studies qualifies its conclusions by noting the several conceptual and data limitations associated with an aggregate quantification of the returns to the economy of research and development investment. (From the standpoint of the study for this book, a major limitation is the assumption that each dollar of NASA research and development spending is equal.) No distinction is made, for example, between spending funds to promote the development of new electronic components and the funding of new instrumentation techniques to support astronomical experiments. There is no illumination of how benefits come about or how institutions change.

A unique and provocative exploration of NASA is the American Academy of Arts and Sciences study of the social impact of the space program initiated in 1962. Because the Academy working group believed that second-order social consequences of actions are often more important than the original action, the project was directed toward anticipating and detecting the second-order effects of massive technological innovation. The study group felt that anticipation or early detection of second-order consequences would affect the choice of a program's goals or the manner in which they are pursued. If undesirable consequences of the space program could be detected during its early years, action could be taken to avoid or offset these consequences. Alternatively, the early anticipation of desirable second-order effects would enable program managers to facilitate these potential benefits (Bauer, 1969).

An integral part of the Academy's study was the application of the historical analogy of the American railroad industry to the U.S. space program. The long-range technological, managerial, and social consequences of the nation's efforts to span the continent with railroads were explored in depth. A historical analogy was then used to determine if the long-range consequences of this earlier nineteenth century innovation suggested the potential second-order effects that might result from space exploration (Mazlish, 1965). This approach provided an interesting analysis of the railroad industry, but it proved less useful in assessing space efforts because so little time had passed since the program began. Historical analysis could work only after more historical evidence had been accumulated.

NASA's Influence on Local Communities

The acceleration of manned space flight programs, which began in 1961, resulted in a significant expansion of existing federal facilities in Florida and Alabama, and the establishment of three entirely new NASA facilities in Mississippi, Louisiana, and Texas. In the early 1960s, NASA established the Marshall Space Center in Huntsville, Alabama; the Kennedy Space Center in Brevard County, Florida; the Manned Spacecraft Center in Houston, Texas; the Michoud Assembly Facility in New Orleans, Louisiana; and the Mississippi Test Facility in Hancock County, Mississippi. Because NASA's manned space flight activities have been concentrated in a "southern crescent" along the Gulf of Mexico, the space program has had a notable influence on this region.

Economic Impacts

The space program's economic impacts in the South have received substantial attention. Earlier studies (Konkel and Holman, 1967; Konkel, 1968; Hough, 1968; and U.S. Congress, 1964) have generated a number of propositions regarding the regional economic effects of large public programs. We shall examine three of them here.

Proposition one: The economic impact of NASA in a locality varied directly with the importance of program employment in the community.

The significance of space employment varied greatly among the five manned space flight communities; NASA's share of local employment was far greater in Hancock County (Mississippi), Brevard County (Florida), and Huntsville (Alabama), in that order, than in either Houston or New Orleans. Specifically, NASA employment, both civil service and contractor, comprised 57 percent of the total 1966 employment in Hancock County, 22 percent in Brevard County, 17 percent in Huntsville, 3 percent in New Orleans, and less than 2 percent in Houston.

In those areas where NASA accounted for a large proportion of local employment, the economic impacts of the space program were direct and identifiable. Hancock County, Brevard County, and Huntsville each experienced large increases in sales volume of local business establishments and growth in per capita income (Table 1-1). Between 1960 and 1965, average increases in retail sales volume and per capita income were 39 and 86 percent respectively for the three communities.

Proposition two: The economic impact of NASA varied inversely with the strength of the local economy at the inception of the program.

A comparison of NASA's economic impacts on Houston with those of New Orleans illustrates this second proposition. The two cities represented strikingly dissimilar economic environments prior to the NASA buildup. New Orleans, which is the much older city, and Houston both achieved population levels of four hundred thousand during the early 1940s. By 1968, Houston's population was two million people, a level almost twice that of New Orleans. Houston's population increase during the ten-year period before 1968 matched the total population growth in New Orleans between 1928 and 1968.

Employment growth in Houston had been sustained at a very high rate since 1940. The annual growth rate during the 1950s was 4.2 percent as compared to a national rate of employment growth of 2.2 percent. In contrast, the annual rate of employment growth in New Orleans during the 1950s was 1.7 percent. The 1957-58 economic recession had a much more severe impact on New Orleans. Unlike the rest of the nation, which had recovered from the 1957-58 recession by 1959, total employment in

Table 1-1

Program Impact on Huntsville, on Brevard County, and on Hancock and Pearl Counties (1960 and 1965)

Category	Population		School Enrollment		Public School Classrooms		Residential Building Permits		Personal Income per Capita		Retail Sales	
	Number[a]	Percent	Number[a]	Percent	Number	Percent	Number	Percent	Income	Percent	Amount[b]	Percent
Huntsville:												
1960	72.0		15.3		568		1,436		$1,537		$111.3	
1965	144.0		32.2		1,010		5,066		2,054		207.8	
Percent		100%		110%		78%		236%		34%		87%
Brevard County:												
1960	111.0		20.2		651		2,614		2,319		125.4	
1965	225.0		48.2		1,514		6,933		3,435		291.3	
Percent change		103		137		133		165		48		132
Hancock and Pearl River counties:												
1960	36.4		8.7		NA		NA		1,134[c]		20.5[d]	
1965	42.3		10.2		NA		NA		1,528[c]		49.3	
Percent change		16		17		—		—		35		40

NA = not available.
[a] In thousands.
[b] Millions of dollars.
[c] Average for Hancock and Pearl River counties.
[d] Retail sales for 1961.

Source: Ronald Konkel and Mary Holman, *Economic Impact of the Manned Space Flight Program* (Washington, D.C.: National Aeronautics and Space Administration, January 1967), chapters 4, 5, and 7.

New Orleans did not regain its 1957 peak of about 292,000 until 1963.

Between 1961 and 1966, employment at both the Michoud Assembly Facility (New Orleans) and the Manned Spacecraft Center (Houston) increased by about eleven thousand personnel. Although the employment increases were roughly similar, the economic impact on the depressed New Orleans economy was far greater. New Orleans, which had experienced an increase in the rate of unemployment from 2.7 percent in 1957 to 6.2 percent in 1961, became one of the ten fastest growing cities in the nation between 1961 and 1966. Space employment directly accounted for 17 percent of the total increase in wage and salary employment during this period. If the secondary employment gain of NASA is included, 38 percent of the total employment growth in New Orleans during this period was due to the space program. It is estimated that New Orleans' employment growth between 1961 and 1966 was 60 percent higher than it would have been without NASA.

In comparison, NASA employment directly accounted for 10 percent of total employment growth in Houston between 1961 and 1966. If secondary effects are included, 30 percent of total employment growth may be attributed to the Manned Space-craft Center. Houston benefited less, relatively, from space employment than did New Orleans; the employment growth was 40 percent higher with NASA than it would have been without the space program influence (the comparable figure for New Orleans was 60 percent).

Proposition three: There are substantial differences in the capabilities of individual communities to diversify beyond the dominant federal program and build an economic base for longer term growth. These differences in capabilities are due to the nature of the federal program on the one hand and local geographic, political, and social characteristics on the other.

The economic growth generated by government programs works in both directions; the local area also becomes more vulnerable to cutbacks or complete elimination of federal pro-grams. The contrasting experiences of Huntsville and Brevard

County illustrate that all communities are not equally vulnerable to contraction of the one large government program.

The city of Huntsville was far more successful than Brevard County in diversifying beyond the dominant NASA program. While much of Huntsville's success was due to organized industrial development, the technological characteristics of NASA activities at the Marshall Space Flight Center afforded Huntsville an important advantage in its diversification efforts. The primary responsibility of Marshall was the manufacturing and testing of rocket propulsion units, such as the first and second stages of the Saturn V launch vehicle. In contrast, the John F. Kennedy Space Center at Cape Kennedy was NASA's prime launch facility and, as such, involved no development or manufacturing activities. The engineering and manufacturing programs at Marshall provided a firmer base for attracting industry than did the launch, maintenance, and technical service activities at Cape Kennedy.

Huntsville capitalized on the industrial opportunities at the Marshall Space Flight Center and the Redstone Arsenal by developing an active and well-planned campaign to attract industry and educational institutions to the city. The Huntsville Industrial Expansion Committee was instrumental in raising funds and mobilizing support for the establishment of the University of Alabama Research Institute. In 1960, Wernher von Braun, director of the Marshall Space Flight Center, assisted the expansion committee by making an appeal before the Alabama state legislature for funds to establish the Institute. A state bond issue provided $3 million for buildings and equipment, and an additional $400,000 was pledged by the city of Huntsville and Madison County. A concurrent development was the creation of Industrial Research Park by a nonprofit group known as Research Sites Foundation, Inc., a land-holding arm of the industrial expansion committee. This organization leased and sold properties on a 200-acre tract adjacent to the arsenal to private firms and research groups at attractive rates.

As part of the plan to attract more industry, particularly non-space-related companies, Huntsville undertook new programs to improve the quality of public services; the local government replaced public buildings, such as the city hall and public library, and established a new municipal airport. The Huntsville Jet Port was identified as a major factor in the decision of nonspace firms, such as Dunlop Tire, Pittsburgh Plate Glass, American Electric, and Barber and Colman, to locate in Huntsville.

As the statistics in Table 1-1 document, the economic changes in Brevard County — higher per capita incomes, more schools, increased construction activity, and so on — are comparable to those in Huntsville. Unlike Huntsville, however, Brevard County made little progress toward sustaining longer term industrial growth and economic stability. As noted earlier, launch activities at Cape Kennedy did not provide a firm basis for attracting new nonspace firms. These limitations were accentuated by the geography of Brevard County and the individualistic nature of major communities in the county.

The geography of Brevard County separates the major population centers in way which inhibits cooperative effort among communities. The distance from Titusville, the largest northernmost town, to Melbourne and Eau Gallie, the major population centers in the south, is about forty miles; the entire area is situated on two thin strips of land, one isolated from the other by the intercoastal waterway. Towns along the mainland are themselves isolated from one another as well as from the beach towns, and the NASA facilities location on Merritt Island provides an additional separation factor. As a result, there is no central city in Brevard County; instead, there are four cities almost equal in population which provide a center for their particular area in the county — Titusville in the north, Cocoa in the center, and Eau Gallie and Melbourne in the south.

The geographical separation of Brevard County communities has been compounded by the desire of each town to preserve its individuality. Despite great pressures since 1950 for cooperation to attack common problems, the communities have never chosen to combine their resources. Smaller cities have developed in each of the population centers, in some cases almost surrounding the larger, older cities. The resulting fragmentation of political authority among the separate cities made it difficult to achieve the concerted industrial development which has characterized Huntsville.

Social Impacts

In addition to direct economic impacts, NASA has altered the quality and context of the local environment in the southern crescent. The influx of large numbers of scientists, engineers, and other professional personnel to these small cities led to the expansion of university and graduate programs in the local area.

The expansion of university programs in Huntsville, Alabama, and southwest Mississippi illustrates the influence of federal research and development activities. The Huntsville Center of the University of Alabama and the university's Research Institute developed concurrently with the growth of NASA's Marshall Space Center and the Army's Redstone Arsenal complex. During the mid-1960s, enrollment in the University of Alabama at Huntsville grew to more than four thousand students, with more than fifteen hundred estimated to be dependents of federal employees at NASA, Redstone, and employees of related industries. This four thousand plus enrollment compared with a total enrollment of fifteen hundred in degree-granting institutions in the Huntsville area prior to 1958. The government's research and development activities provided the impetus for the University of Alabama to become the largest graduate engineering school in the South as of 1964.

The NASA Mississippi Test Facility had a similar influence in upgrading education facilities throughout southwest Mississippi. In September 1965, the University of Southern Mississippi established an extension center offering undergraduate courses in the city of Picayune, which is located in Hancock County. Pearl River Junior College began to offer evening courses in Picayune to equip students with the specialized skills required for positions at the local NASA facility. In the fall of 1965, the Jefferson Davis Junior College, which is located in Harrison County, Mississippi, about midway between Gulfport and Biloxi, began operation.

The educational impact of federal research and development programs was not limited to the university and junior college level; primary and secondary school systems also improved markedly. The huge increases in populations and school enrollments in southern communities are shown in Table 1-1. Moreover, the rapid growth in school enrollments and construction were accompanied by substantial improvements in average education attainment and in educational quality at the primary and secondary levels. In Huntsville, the 1950 figure for median school years was well below the national average (7.5 as against 9.3 years); by 1960, the Huntsville average was 0.2 years above the then national average of 10.6 years. By 1964, between 75 and 80 percent of all Huntsville secondary school graduates entered college.

The composition of the educational staff and course content also changed substantially. In 1964, some 350 spouses of NASA and Redstone personnel or of local defense-related industry employees served as teachers in the Huntsville system. Of the more than eight hundred teachers in the city school system, approximately one-fourth held masters' degrees, a high percentage compared with other Alabama cities. Significant improvements in curriculum content were made, with advanced biology and calculus added to the secondary school programs.

Another educational innovation for the Huntsville area was the establishment of an extension unit of the state vocational-technical school in the early 1960s. The extension unit offered high school students the opportunity to complete the eleventh and twelfth grades with training in electronics, auto mechanics, and related technical fields. The effect of this new program was to upgrade substantially the labor force for the entire area. Many of the graduates of the technical school found ready employment at the Redstone Arsenal and the Marshall Space Center.

NASA's Influence on Industry

The third major area which previous NASA studies have explored is the program's impact on private industry. Specifically, these studies have dealt with NASA's impact on market demand, industrial technology, and management practices.

Market Demand

As a result of NASA's multibillion-dollar budget during the 1960s, the agency became an important customer for several U.S. industries. An input-output analysis of NASA's expenditures for fiscal years 1966 and 1967 provided an industrial breakdown of the agency's direct expenditures for goods and services. The industrial breakdown, based on the four-digit standard industrial classification (SIC) code, revealed that NASA expenditures were distributed in more than one hundred different industries (Orr and James, 1969). The industries with the largest NASA expenitures were aircraft (SIC 3721), aircraft equipment (SIC 3729), and complete guided missiles, system assembly, and related engineering (SIC 3722).

The space program expenditures had a significant influence on the U.S. aerospace industry. Total sales of aerospace products increased from $16.4 million in 1961 to $22.6 million in 1967. Among the three major components of industry demand — Defense Department, NASA and the Atomic Energy Commission, and commercial purchases — there was a significant shift in relative importance during this period. The combined NASA and Atomic Energy Commission share of total industry demand rose from 4 to 19 percent, commercial purchases from 11 to 18 percent, and the Defense Department demand dropped from 85 to 63 percent (Table 1-2). The impact of the manned space flight program on the aerospace industry is based on the study by Konkel and Holman (1967, pp. 31-70).

The importance of NASA as a source of industry demand increased each year from 1960 to 1965. From 1963 to 1965, the rising demand from NASA programs served to offset decreasing defense purchases within the industry. The major impact of increased NASA purchases in the aerospace industry during this period was a shift of employment away from aircraft production into missiles and space production.

In 1966, the relative importance of NASA purchases began to decline because of large increases in defense and commercial purchases. Demand for military aircraft rose as a result of the U.S. commitment in Vietnam; there was also a rise in the commercial demand for transports. As a result, the aerospace employment shift of 1960-65 was reversed. By 1966, employment in missiles and space had declined by about seventy thousand from its peak level of 578,000 in 1963.

Industrial Technology

New technology provided by a government program can substantially alter the matrix of products, processes, techniques, methods, and materials within an industry. A dramatic example of such a technological impact is the influence of NASA's communications satellites on the growth of the international communications industry. The distinctive contribution of these satellites is best understood by tracing the historical development

Table 1-2
The Structure of Aerospace Demand [a]
(1961-67)

Year	Total Aerospace Products and Services	Government Purchases		Department of Defense		NASA and Atomic Energy Commission		Commercial Purchases	
		Amount	Percent[b]	Amount	Percent[b]	Amount	Percent[b]	Amount	Percent[b]
1961	$16,377	$14,501	88.5%	$13,871	84.7%	$ 630	3.8%	$1,876	11.5%
1962	17,437	15,665	89.8	14,331	82.2	1,334	7.6	1,772	10.2
1963	18,304	16,819	91.9	14,191	77.5	2,628	14.4	1,485	8.1
1964	18,873	16,853	89.3	13,218	70.0	3,635	19.3	2,020	10.7
1965	18,702	15,886	84.9	11,396	60.9	4,490	24.0	2,816	15.1
1966	21,350	17,950	84.1	13,110	61.4	4,840	22.7	3,400	15.9
1967[c]	22,550	18,550	82.3	14,150	62.8	4,400	19.5	4,000	17.7

[a] Millions of dollars.
[b] Percentage of total.
[c] Estimated.

Source: Aerospace Industries Association. *Aerospace News* (December 20, 1966). Table 1.

of this industry (Martin, 1969; Wilson, 1953; International Communications Union, 1965). International communications date back to 1866 when the British installed the first transatlantic ocean cable. After World War I, transoceanic communication via high-frequency radio developed into an industry with RCA the leading firm.

The first transatlantic telephone cable (TAT-1) was laid in 1956. The cable overcame many of the transmission difficulties which characterized high-frequency radio. Due to the success of the cable, international communication services, particularly telephone, experienced exceptional growth. Three additional cables (TAT-2, -3, and -4) with improved design and capacity linked the United States and Europe by 1965. The cables were originally laid by the American Telephone and Telegraph Company, with associated foreign countries sharing in their ownership. Although they were primarily intended for telephone use, they were easily adapted to handle all varieties of telegraph or record traffic and soon led to the obsolescence of the earlier (pre-1956) cables.

The first satellite communications system, INTELSAT-I, became operational in 1965 after a fifteen-year national research and development effort to build commercial communications satellites. The first effort in satellite communications was the Navy Communication by Moon Relay project in which radar signals and voice communications were transmitted from Pearl Harbor to Washington, D.C., via the moon. The first communications program of NASA was Project Echo, which originated with NASA's predecessor, the National Advisory Committee for Aeronautics, in 1956. Echo I, which was a hundred-foot aluminum plastic balloon, was successfully launched into a thousand mile orbit on August 12, 1960. Because Echo influenced subsequent technology in ground stations and radio propagation, and publicly demonstrated the concept of satellite communications, it was a historic milestone. During the 1958-63 period NASA, with the assistance of various military agencies, AT&T, Space Technology Laboratories, and Hughes Aircraft, carried out several additional innovations to advance satellite communications technology. (The role of NASA in the development of communications satellites is discussed in detail in the study by the Midwest Research Institute, November 1971.)

The Communications Satellite Act of 1972 created COMSAT to manage the country's commercial communications satellite system. It (COMSAT) operates as a carrier's carrier and leases satellite circuits to the International Record Carriers, who in turn lease circuits to the public. Because of their greater channel capacity, INTELSAT-I spurred a major expansion in the volume of international communications. Between 1966 and 1970, the volume of international communications (excluding telephone) increased by 55 percent; this increase represented the highest rate of growth for any five-year period since 1961. By 1971, COMSAT had invested almost $200 million in equipment and facilities; between 1965 and 1970, COMSAT's revenue grew from just over $2 million to nearly $70 million (Interaglia, 1972, p. 52).

The new technology which industry acquired from NASA had a significant effect on the industry's cost structure. The annual cost of a satellite communications circuit was $25,000 when INTELSAT-I was launched in 1965; the cost had dropped to $719 when INTELSAT-IV was launched in 1971. It is estimated that the annual cost of a circuit will drop to $30 by 1976 when INTELSAT-V will be placed in orbit. The cost of earth stations has also declined substantially; whereas space station costs ranged between $6 million to $12 million in 1968, the range had been reduced to between $2 million and $4.5 million by 1971 (Midwest Research Institute, 1971, pp. 54-59). The commercial results of these technological advances are reflected in the recent history of transatlantic telephone charges. For example, the monthly rates for a leased telephone circuit between New York and Paris remained unchanged for a number of years; but in 1966, immediately after the first communications satellite went into operation, monthly charges dropped sharply and have continued to drop since that time. The $4,625 monthly charge in 1971 was less than half the monthly rate for 1965 (Jastrow and Newell, 1972, p. 541).

Technological Progress

In addition to generating specific technological gains for industry, NASA has also sought to promote general technological progress. In 1962, NASA became the first federal agency to establish an agencywide program to promote the transfer and application of its technology to outside agencies. The 1958 National Aeronautics and Space Act provided a statutory basis for the NASA Technology Utilization Program by calling for the widest possible dissemination of information concerning space activities and for the study of the long-range benefits to be derived from space.

When the Technology Utilization Program was created, it was widely believed that the technical by-products of the space program, or "space spinoffs," would be both large in number and commercially significant. The concept of "space spinoff" assumed that a specific, discrete innovation in the space program would be identified as relevant to a need outside the program. It would then be adapted and applied commercially. (The history and operation of NASA's Technology Utilization Program are discussed in Lesher and Howick, 1966; Doctors, 1968, p. 69; Rosenbloom, 1969, pp. 156-65; and Geise, 1971.)

The first substantive study of space spinoffs was commissioned by NASA and carried out by the Denver Research Institute in 1963. The Institute's researchers were unable to identify a significant number of technical by-products. Moreover, they suggested that the term "spinoff" was misleading because it implied that space contributions were direct and readily identifiable, when in fact they were not (Denver Research Institute, 1963; Welles and Waterman, 1964, p. 106).

Since 1963, a more realistic view of the space program's technological influences has been adopted. The Technology Utilization Program is now focused on developing improved methods of technology transfer, rather than generating "space spinoffs." This new approach was assessed in a 1972 study of NASA's technological influences on industry (Denver Research Institute, May 1972).

The 1972 study proceeded as follows: Non-NASA technical experts in industrial firms, universities, and government laboratories were asked to identify the major developments in twelve technological fields — cryogenics, electrochemical energy conversion and storage, high-temperature ceramics, high-temperature metals, integrated circuits, internal gas dynamics, materials machinery and forming, materials joining processes, microwave systems, nondestructive testing, simulation, and telemetry. A total of 109 developments were identified in this fashion.

The technical people of NASA then specified 366 NASA contributions to the advancement of the 109 selected developments.

The contributions of NASA per field ranged from a low of fourteen in telemetry to a high of 61 in high-temperature metals, with an overall average of approximately thirty NASA contributions per field. The NASA respondents assessed the nature, significance, and impact of the space program's contributions and also the magnitude of the technological change represented by these contributions.

The most significant finding of the 1972 study is that the principal technological impact of the space program is to accelerate technical advance. The survey indicates that a major share (78 percent) of NASA's contributions were advances that would have eventually occurred even in the absence of the space program; NASA's role was to accelerate development. The study also found that most NASA contributions (65 percent) represented an incremental or systematic advance in technology rather than a more significant "step function" change or a less significant "consolidation of knowledge."

Mathematica, Inc., has recently conducted a "pilot study" to measure the economic value of NASA's acceleration effect (June 1975). Drawing on four of the Denver Research Institute's case studies (gas turbines, cryogenic multilayer insulation, computer simulation, and integrated circuits), Mathematica, Inc., concluded that the economic benefits which result from NASA's acceleration of technology are indeed large. The value of a speed-up in technology in those four fields was estimated to be between $2.3 billion and $7.6 billion (in 1974 dollars). Mathematica, Inc.'s most probable estimate is that the four cases in the pilot study alone produced savings equal to 6 percent of all NASA research and development expenditures since 1958 (on a discounted value basis).

Management Impacts

Improvements in management techniques as well as technological advances have been viewed as an impact of NASA. A 1971 Denver Research Institute study was the first systematic attempt to examine the application of aerospace management techniques outside the aerospace industry. A total of 25 techniques, used widely within the aerospace industry, were identified. The distinguishing characteristic of the techniques is *use* and not *origination* in aerospace. Many of the 25 techniques (such as program and review techniques (PERT), simulation analysis, and management information systems) originated elsewhere; aerospace managers have played an important role in refining, transforming, or augmenting the techniques.

The principal contribution of the Denver Research Institute's study is in furthering understanding of the processes for transferring management technology. The research team concluded that the transfer and application of aerospace management technology occur primarily through the movement of the industry's trained and experienced people to organizations outside the aerospace field.

Another recent study of NASA's management contributions was conducted by Sayles and Chandler (1971), under the auspices of the National Academy of Sciences. The authors focused on what future large-scale programs can learn from NASA. They conclude that NASA's most outstanding contribution is in the area of advanced systems design: ". . . getting an organizationally complex structure, involving a great variety of people doing a great variety of things in many separated locations, to do what you want, when you want it" (p. 314).

The distinctive skill of NASA in advanced systems design was its ability to integrate widely separated, specialized efforts. Space program managers employed a variety of methods (such as data banks, configuration management, and endless performance review meetings) to attain what Sayles and Chandler term "systemwide performance visibility." The internal management control capabilities of NASA enabled the space agency to penetrate its total work system in an almost unprecedented manner. Although Sayles and Chandler clearly document NASA's integrative management control skills, their study does not indicate the extent to which these management skills have been adopted by other large organizations, or the blocks to their adaptation.

Summary

Previous NASA studies have explored the economic and social impacts of the space program on southern communities, the impact of NASA on industrial demand, technology, and management, and the relationship of NASA to the social economy.

First, space employment had a major impact on the economies of the five NASA manned space flight communities bordering the Gulf of Mexico — the "southern crescent." The economic impact of NASA on a local community varied directly with the relative importance of program employment in the community and inversely with the strength of the local economy at the inception of the program. Diversification of the economy beyond the dominant federal program varied with the nature of the particular space facility and the individual community's geographic and political characteristics.

Second, the NASA experience demonstrates that high-technology public programs can have an important influence on the quality and context of a local environment. These social impacts of government research and development programs are especially noticeable in smaller localities which have a limited -technology base. In such communities the influx of scientists, engineers, and professional personnel helped to produce marked improvements in the primary and secondary school systems. In addition, new junior college and university facilities were established and existing institutions were greatly expanded.

Third, government research and development programs can provide important new technology for particular industrial sectors, as NASA did for the international communications industry. Indeed, NASA was the leader in a fifteen-year research and development effort which produced INTELSAT-I, the first operational communications satellite system. The greater channel capacity of satellites stimulated a 55 percent expansion in the volume of international communications (excluding telephone) between 1966 and 1971. The continual decline in the costs of satellite circuits has been accompanied by reductions in the prices of all forms of international communication; 1971 transatlantic telephone rates, for example, were almost half those of 1965.

Fourth, the space program's primary technological influence has been to accelerate technical development. A study of more than a hundred developments in a variety of technical fields suggests that most technical advances occurred earlier than would have been expected in the absence of NASA. These findings suggest that earlier "space spinoff" studies, which searched for commercial adoption of complete NASA-developed systems, were misdirected.

Fifth, NASA's most significant contribution to management practice is in the design of advanced systems. The space program developed and refined management methods for achieving objectives in extremely large and complex systems. At the present time, the adoption of NASA's management techniques by organizations in the private sector has proved difficult to identify and document.

A more general conclusion is that earlier studies have tended to deal with only one category of impacts — economic, social, technological, or managerial. This approach has focused attention on the type of impact rather than on the affected institution; most earlier impact studies have failed to recognize the existence of and potential for institutional transformations. Yet the real significance of public program impacts is that they may ultimately serve to modify and transform institutional arrangements. In order to overcome the limitations of past studies, the Conservation project undertook three case studies organized round an institutional framework.

References

Bauer, R. Editor. *Second-Order Consequences.* Cambridge, Massachusetts: M.I.T. Press. 1969.

Benoit, Emile; and Boulding, Kenneth E. Editors. *Disarmament and the Economy.* New York: Harper & Row, Publishers, Incorporated. 1963.

Chase Econometric Associates, Inc. "The Economic Impact of NASA R&D Spending: Preliminary Executive Summary." Bala Cynwyd, Pennsylvania: Chase Econometric Associates, Inc. April 1975.

————. "Relative Impact of NASA Expenditure on the Economy." Bala Cynwyd, Pennsylvania: Chase Econometric Associates, Inc. March 18, 1975. Unpublished staff report.

Denver Research Institute. *Aerospace Management Techniques: Commercial and Governmental Applications.* Prepared for NASA. Denver, Colorado: Denver Research Institute. November 1971.

_____. *The Commercial Application of Missile/Space Technology.* Denver, Colorado: Denver Research Institute. September 1963.

_____. *Mission-Oriented R&D and the Advancement of Technology: The Impact of NASA Contributions.* Denver, Colorado: Denver Research Institute. May 1972.

Doctors, S. *The Role of Federal Agencies in Technology Transfer.* Cambridge, Massachusetts: M.I.T. Press. 1968.

Geise, J. *The Role of the Regional Dissemination Centers in NASA's Technology Utilization Program.* Washington, D.C.: National Aeronautics and Space Administration. May 1971.

Hough, Roger W. *Some Major Impacts of the National Space Program. Part V, Economic Impacts.* Menlo Park, California: Stanford Research Institute. June 1968.

Interaglia, Dominick A. "The U.S. International Record Carrier: Past, Present, and Future." New York: Pace College. February 1972. Master's thesis.

International Telecommunications Union. *From Semaphore to Satellite.* Geneva: International Telecommunications Union. 1965.

Jastrow, Robert; and Newell, Homer E. "The Space Program and the National Interest." *Foreign Affairs* (April 1972).

Konkel, Ronald. "Space Employment and Economic Growth in Houston and New Orleans, 1961-1966." New Orleans, Louisiana: Tulane University. June 1968. Master's thesis.

_____; and Holman, Mary. *Economic Impact of the Manned Space Flight Program.* Washington, D.C.: National Aeronautics and Space Administration. January 1967.

Lesher, R.; and Howick, G. *Assessing Technology Transfer.* Washington, D.C.: National Aeronautics and Space Administration. 1966.

Martin, James. *Telecommunications and the Computer.* Englewood Cliffs, New Jersey: Prentice-Hall, Inc. 1969.

Mathematica, Inc. "Quantifying the Benefits to the National Economy from Secondary Application of NASA Technology." Princeton, New Jersey: Mathematica, Inc. June 1975.

Mazlish, B. Editor. *The Railroad and the Space Program: An Exploration in Historical Analogy.* Cambridge, Massachusetts: M.I.T. Press. 1965.

Midwest Research Institute. "Technological Progress and Commercialization of Communications Satellites." In *Economic Impact of Stimulated Technological Activity.* Kansas City, Missouri: Midwest Research Institute. November 1971.

Oliver, Richard P. "Employment Effects of Reduced Defense Spending." *Monthly Labor Review* (December 1971).

Orr, Lloyd; and James David. "An Industrial Breakdown of NASA Expenditures." Bloomington: Indiana University. November 1969.

Rockwell International, Space Division. "Impact of the Space Shuttle Program on the California Economy." Pittsburgh, Pennsylvania: Rockwell International. December 1974.

Rosenbloom, R. "The Transfer of Space Technology." In *Second-Order Consequences.* Edited by R. Bauer. Cambridge, Massachusetts: M.I.T. Press. 1969.

Sayles, L.; and Chandler, M. *Managing Large Systems: Organizations for the Future.* New York: Harper & Row, Publishers, Incorporated. 1971.

Solow, Robert A. "Gearing Military R&D to Economic Growth." *Harvard Business Review* (November-December, 1961-62), vol. 40, no. 6.

U.S. Congress, House. *Impact of Federal Research and Development Programs.* 88th Cong., 2d Sess. Washington, D.C.: U.S. Government Printing Office. 1964.

U.S. Department of Labor, Bureau of Labor Statistics. *Manpower Impact of Federal Government Programs: Selected Grants-in-Aid to State and Local Governments.* Report 424. Washington, D.C.: U.S. Department of Labor. October 1973.

Welles, J. G.; and Waterman, R. H. "Space Technology: Pay-Off from Spin-Off." *Harvard Business Review* (July-August 1964).

Wilson, George L. *Transportation and Communications.* New York: Appleton-Century-Crofts, Inc. 1953.

Chapter Two

New Industries: Public Support for Semiconductor and Computer Development

Electronics has been one of the world's fastest growing industries during the twentieth century, expanding almost a hundredfold between 1939 and 1974. The industry's most rapid expansion has occurred since the mid-1950s as the result of dramatic growth in world computer and semiconductor markets. The early development and subsequent growth of these two sectors of the electronics industry have been due to the significant technological advances since World War II.

American firms have dominated the semiconductor (solid-state electronic components) and computer industries since their post-World War II origins. International Business Machines, for example, accounts for more than two-thirds of all computer installations both inside and outside the United States. In semiconductors, American firms have carried out all of the notable innovations since Bell Laboratories discovered the transistor in 1948 (Freeman, 1965, p. 1).

Large U.S. government programs in national defense and space exploration have played a major role in the development of the nation's computer and semiconductor industries. This chapter identifies and describes three types of influence — economic, technological, and manpower — which the space and defense programs exerted on these new industries. A comprehensive analysis of all three impact categories is needed to understand the full range of support a public program can provide to a new industry. The influence of government policy and programs on the electronics industry is not a post-World War II phenomenon; some segments of this industry (e.g., radar and equipment for radio broadcasting) have had a continuing involvement with government policy since World War I (Freeman, pp. 52-56; Postan, Hay, and Scott, 1964; and MacLaurin, 1949).

Economic Impacts: Support for Research and Development

The economic dimension of government influence has been evident in both the computer and the semiconductor industries. Specifically, three types of economic impacts can be identified:

(1) Direct and indirect financial support by space and defense programs for semiconductor and computer research and development

(2) Assured demand provided by space and defense programs during the early years of the industry

(3) The use of space and defense demand to support new firms and to affect the competitive balance within the industry as it matures

Because innovation — the application of technology — has been so crucial in the development of the computer and semiconductor industries, let us begin by considering space and defense financial support for research and development.

The development of active (or current modifying) electronic components may be divided into three "generations." The first consisted of electronic tubes which were produced commercially beginning in 1920. The invention of the transistor by Bell Laboratories in 1948 marked the birth of the second generation of active electronic components. Whereas a tube altered electrical current as it passed through a vacuum or inert gas in the tube, a semiconductor could alter current as it passed through solid materials. The use of solid materials increased reliability and reduced the size of components, thereby permitting miniaturized and more complex products. The second generation is familiar to the average consumer in the form of portable transistor radios as well as numerous other products. A third generation of electrical components, integrated circuits, was introduced to the market in the early 1960s. Integrated circuits are combinations of two or more electrical components joined together within a solid-state material (Tilton, 1971, pp. 7-9). The use of integrated circuits permitted "micro-miniaturization," and thus highly complex circuits could be housed in a reasonable space. Integrated circuits have recently reached the consumer in the form of "pocket calculators."

Prior to the discovery of the transistor, European firms were the equal of American firms in advanced component technology. Some of the most significant advances in electron tubes were made by European firms and European universities. The role of European firms in advanced component technology has declined considerably since 1948. Almost all of the important inventions and innovations in components have been made in the United States, and there has been a lag of one to four years before manufacture has begun in Europe, often by American subsidiaries (Freeman, 1965, p. 63).

The leading role that American firms have played in advanced component technology is directly related to the research and development support provided by the government. In the early 1950s, the military services became convinced of the military value of transistors and other semiconductor devices. They believed these devices could miniaturize, and thus make more portable, considerable military equipment and also improve the reliability of equipment in field operations. As a result, the Defense Department financed semiconductor research and development on a large scale and aided the principal firms in enlarging their production capacity. By 1953, the Army Signal Corps was financing pilot production lines for transistors and related devices at five sites operated by Western Electric, General Electric, Raytheon, RCA, and Sylvania (Tilton, 1971, pp. 92-95). Between 1955 and 1961, direct government funding for semiconductor research and development and production refinement totaled $66 million (Table 2-1).

This government financial support was provided in two forms: (1) direct contracts for research and development projects and (2) government-financed production refinement programs. Approximately 30 percent of the funds were expended in 1956 when a special appropriation of $14 million was allocated to production refinement projects for transistors. Contracts for the development of thirty different types of transistors over the following several years were placed with twelve semiconductor companies. In many

Table 2-1

Direct U.S. Government Funding of Semiconductor Research and Development and Production Refinement[a]

(1955-61)

| | | Production Refinement | | |
Year	Research and Development	Transistors	Diodes and Rectifiers	Total
1955	$ 3.2	$ 2.7	$2.2	$ 8.1
1956	4.1	14.0	0.8	18.9
1957	3.8	0.0	0.5	4.3
1958	4.0	1.9	0.2	6.1
1959	6.3	1.0	0.0	7.3
1960	6.8	0.0	1.1	7.9
1961	11.0	1.7	0.8	13.5

[a] Millions of dollars.

Source: Business and Defense Services Administration, U.S. Department of Commerce, *Semiconductors: U.S. Production and Trade* (Washington, D.C.: Government Printing Office, 1961). Table 8.

cases the Defense Department's investment was matched by similar amounts of capital equipment or plant space supplied by the contracting companies. Contracts called for the delivery of about three thousand transistors of each type to be made on production lines capable of producing that many per month. Eventually the special $14 million expenditure aided in greatly expanding semiconductor production capacity in the United States (Arthur D. Little, Inc., 1963, p. 163).

The Army Signal Corps was extremely active in supporting semiconductor research during this period because of its concern with improving military communications equipment. In 1961, this agency employed 27 commercial contractors on research programs for semiconductor devices, forty commercial firms on electronic parts and materials projects, and thirteen commercial firms on quartz crystal units. The Signal Corps program produced some notable commercial benefits for these firms by enabling them to substantially reduce the length of commercial development projects. The Signal Corps believes that the time needed on these projects was shortened by as much as 75 percent (U.S. Congress, Senate, 1961).

It should be noted that the data on direct government funding of semiconductor research and development (Table 2-1) vastly understate the amount of total government support. Many semiconductor firms received support indirectly. Government contracts for space and weapons systems also enabled the prime contractors to subcontract with semiconductor firms for research and development. A 1960 Defense Department survey measured the amount of direct federal semiconductor support as well as the indirect amount which was subcontracted to semiconductor firms by weapons systems prime contractors. The survey estimated that Defense Department direct and indirect semiconductor research and development funding totaled $13.9 million in 1958 and $16.2 million in 1959. These totals were over twice the amount of direct Defense Department research and development funding in those years. The survey also revealed the relative importance of government-sponsored research and development in 1958 and 1959; the Defense Department funded 25 percent and 23 percent respectively of total semiconductor industry research and development expenditures (Tilton, 1971, pp. 93-94). As these figures make clear, substantial support was provided for semiconductor research and development during the early stages of the industry's development. The $66 million in direct support, as well as nearly equal amounts in indirect support, aided the firms in refining their products and expanding production capacity.

A similar pattern of early large-scale government research and development support is evident in the computer industry. Federal agencies began to play an instrumental role in development of the U.S. computer industry in the 1940s. Although the first U.S. computer, the automatic sequence controlled calculator, was developed by a Harvard-International Business Machines team between 1937 and 1944, all subsequent U.S. computer developments from the end of World War II until the mid-1950s had the strong financial support of various U.S. government agencies. Work on the first electronic computer, the electronic numerical integrator and calculator, began at the University of Pennsylvania in 1942 and was completed in 1946. This electronic computer received financial support from the Army and was mainly designed for calculating trajectories of shells and bombs.

From 1945 to 1955, the U.S. Army, Navy, and Air Force, the Atomic Energy Commission, and the National Bureau of Standards all placed major contracts for the development of improved computers with several universities and with the firms who first began design and manufacture — especially Remington Rand (Univac) and, later, IBM. As a result, rapid progress was made in solving some of the problems of logic design, memory storage systems, and programming techniques.

The major technological advances during this period were made in the large computers demanded by the military and other government agencies. These advances were subsequently incorporated in the medium and small computers, which were designed for smaller scale, less complex applications. In the 1950s, government contract research and development accounted for about 60 percent of total research and development expenditures at IBM; by the early 1960s, government-sponsored research and development was but a small part of the company's $125 million research and development budget. The large IBM computers, such as STRETCH, were built on contract for government agencies, but the highly successful transistorized 1401 series and the 360 and 370 series have been privately funded projects (Burck *et al.,* 1965, p. 82).

Government support of U.S. computer developments takes on added significance when the U.S. experience is contrasted with that of other countries. Although it is widely believed that the first successful computers were built in the United States and Great Britain, the first electromechanical computer was actually developed by Frederich Zuse in Berlin between 1936 and 1941. Moreover, Zuse was also the first to begin work on the electronic computer during the early 1940s. But the lack of official support by the German government for his project, and the disruption which occurred at the end of the war, caused the lead in computer development to pass to the United States, and to a lesser extent, to Great Britain. In England, electronic computer development work began in 1946-47 after British scientists and engineers had visited the first American installations. The earliest work was at London, Manchester, and Cambridge universities, and at the Elliott Brothers Research Laboratory on digital computing for fire control for the Admiralty. Although financial support for several computer projects was provided by the National Research and Development Corporation, total funding was limited to $2.5 million during the 1947-57 period because British military agencies remained unconvinced that there were substantial benefits to be derived. Consequently, British computer firms were never able to compete effectively with their U.S. counterparts. Initially, ten British firms were involved in the development and manufacture of computers. But few of these firms ever developed a computer model with sales exceeding fifty units, and by the mid-1960s the number of British computer manufacturers had declined from ten to three (Bernstein, 1966; Freeman, 1965, pp. 56-62; and Jacobowitz, 1963, p. 97).

Impact of Assured Demand

The preceding discussion makes it clear that government research and development funding provided vital financial support for the early technological development of semiconductors and computers. Government programs were equally important when these products first reached the market; the earliest sales of semiconductors and computers were to military agencies. The discussion in the remainder of this section, unless otherwise noted, is based upon data found in publications of the Business and Defense Services Administration (U.S. Department of Commerce, 1960), the Electronic Industries Association (1969), and John Tilton (1971, pp. 89-92).

The first semiconductors were marketed in the early 1950s. By 1955, the semiconductor market was $40 million, with military agencies accounting for 38 percent of these sales. The space and defense markets for semiconductors grew from $15 million in 1955 to $222 million in 1961. At times, space and defense demands accounted for almost 50 percent of the total semiconductor market (Table 2-2).

Government demand proved even more significant during the first years of integrated circuit commercial production. The earliest integrated circuits were used in the Minuteman missile program and various Navy programs in 1962 and 1963. In 1962, space and defense sales accounted for the total $4 million integrated circuit market. By 1968, the space-defense share of integrated circuits had declined to 37 percent (Table 2-2).

Space-defense business was vital in an industry where learning economies are substantial. Typically, the price for each new com-

Table 2-2

U.S. Production of Semiconductors and Integrated Circuits for Defense Requirements (1955-68)[a]

Year (1)	Semiconductor Production			Integrated Circuit Production[b]			Semiconductor and Integrated Circuit Production		
	Total (2)	Space and Defense (3)	(3) as a Percentage of (2) (4)	Total (5)	Space and Defense (6)	(6) as a Percentage of (5) (7)	Total (8)	Space and Defense (9)	(9) as a Percentage of (8) (10)
1955	$ 40	$ 15	38%				$ 40	$ 15	38%
1956	90	32	36				90	32	36
1957	151	54	36				151	54	36
1958	210	81	39				210	81	39
1959	396	180	45				396	180	45
1960	542	258	39				542	258	48
1961	565	222	39				565	222	39
1962	571	219	38	$ 4[c]	$ 4	100%[c]	575	223	39
1963	594	196	33	16	15	94[c]	610	211	35
1964	635	157	25	41	35	85[c]	676	192	28
1965	805	190	24	79	57	72	884	247	28
1966	975	219	22	148	79	53	1,123	298	27
1967	879	205	23	228	98	43	1,107	303	27
1968	847	179	21	312	115	37	1,159	294	25

[a] Dollar amounts are in millions. Defense production includes devices produced for equipment needed by the Defense Department, the Atomic Energy Commission, the Central Intelligence Agency, the Federal Aviation Agency, and the National Aeronautics and Space Administration.
[b] Commercial production of integrated circuits began in 1962.
[c] Estimated.

Source: For semiconductors — Business and Defense Services Administration, U.S. Department of Commerce, *Electronic Components: Production and Related Data, 1952-59* (Washington, D.C.: U.S. Government Printing Office, 1960); for 1960-68 — *ibid.,* "Consolidated Tabulation: Shipments of Selected Electronics Components" (annual reports), and Electronics Industries Association, *Electronics Industries Yearbook, 1969* (Washington, D.C.: Electronics Industries Association, 1969). These data appear in a different form in John Tilton, *International Diffusion of Technology: The Case of Semiconductors* (Washington, D.C.: The Brookings Institutions, 1971) Tables 4-7 and 4-8.

Table 2-3
U.S. Integrated Circuit Production and Prices
(1962-68)

Year	Total Production (millions)	Average Price per Integrated Circuit (dollars)
1962	$ 4[a]	$50.00
1963	16	31.00
1964	41	18.50
1965	79	8.33
1966	148	5.05
1967	228	3.32
1968	312	2.33

[a] Estimated.

Source: Total production and average price figures   Electronics Industries Association, *Electronics Industries Yearbook, 1969* (Washington, D.C.: Electronics Industries Association, 1969), Table 55.

ponent innovation was too expensive for industrial or consumer electronic products. But NASA and the Defense Department provided a substantial market for new devices because reliability and performance received higher priority over costs in the space and defense environments. Later, as the firms increased production, learning economies were realized and costs fell. In a few years, the price was low enough to penetrate the industrial market and, eventually, the consumer market.

Semiconductor firms estimate that the average cost of producing a new component falls between 20 and 30 percent every time its cumulative past production doubles. The impact of learning economies on the price of integrated circuits has been dramatic. In 1962, when the total market was $4 million, the average price of an integrated circuit was approximately $50. By 1968, the market had grown to $312 million, and the average price had dropped to $2.33 (Table 2-3). At the 1973 price of $0.63, integrated circuit customers could purchase fifty units of increased complexity for the same price they had paid for one unit in 1963 (Texas Instruments, Inc., 1973, p. 3).

Computer manufacturers, as well as semiconductor firms, benefited from government purchases during the industry's early years. Military demand for computers was critical in the early 1950s. By the time Sperry Rand delivered the first UNIVAC computer designed for business data processing to General Electric in 1954, the military market for data processing systems had already reached nearly $200 million. Although sales of commercial data processing hardware rose rapidly to $500 million in 1960, the military and space market for data processing systems was still larger ($900 million) in that year. The sales of commercial data processing systems did not surpass military and space hardware sales until 1962, nearly eight years after delivery of the first computer designed for business use (Table 2-4).

The relationship between military and space demands for computers and overall market growth is similar to that observed in the semiconductor industry. In both cases, space and defense demands accounted for a large share of the total market at the outset. The space-defense share of the total computer hardware market was initially 100 percent and remained over 60 percent for the next seven years. Although the space-defense share of the computer services market was lower than the space-defense share of the computer hardware market, government purchases still accounted for nearly half of total sales through 1960 amd did not drop below one-third of the total market until 1963 (Burck *et al.,* 1965, pp. 73-78; Bernstein, 1966, pp. 81-82; and Gropelli, 1970).

The above statistics document the direct spending of space and defense agencies on computers, but they do not capture the indirect effect of space and defense expenditures resulting from computer spending by large aerospace companies. A computer science professor of a major eastern university (in a personal interview with the author) described the indirect effects of space and defense programs on the computer industry by saying that "big spenders are significant factors" in the computer industry of America. He noted that American "dominance of computers and computer science" is due in large part to "massive government space-defense spending" during 1950 through 1965, adding that these funds "flowed to large aerospace companies who in turn

Table 2-4
Computer Market Statistics
(1954-63)

Year	Computer Sales[a]			Sales		Space-Defense Share	
	Commercial	Space-Defense	Total	Computer Services	Computers and Services	Computers	Computers and Services
1954		$ 200	$ 200		$ 200	100%	100%
1955	$ 68	250	318	$ 50	368	79	54
1956	185	300	485	175	660	62	45
1957	230	350	580	200	780	60	45
1958	200	500	700	250	950	71	53
1959	310	800	1,110	300	1,410	72	57
1960	600	900	1,500	500	2,000	60	45
1961	820	1,000	1,820	750	2,570	55	39
1962	1,200	1,100	2,300	1,000	3,300	48	33
1963	1,370	1,200	2,570	1,200	3,770	47	32

[a] Millions of dollars.

Source: G. Burck *et al.*, *The Computer Age* (New York: Harper & Row, Publishers, Incorporated, 1965, pp. 73-78; and A. Gropelli, "The Growth Process in the Computer Industry," (New York: New York University, 1970), appendix, Doctoral dissertation.

became big spenders" which resulted in the "emergence of the pioneering IBM 7090 and 7094 lines."

The large space and defense markets for computers produced three significant secondary benefits for computer firms. First, space and defense sales enabled companies to fund higher levels of research and development, which in turn promoted technological progress. Research advances, together with the learning economies which resulted from increasing volume levels, led to continual declines in prices. An index of computer prices for the 1954-65 period reveals that prices in 1965 were approximately a tenth of those in 1954, as the following shows (Chow, 1967, p. 1124; 1960 equals 100):

Year	Computer Price Index
1954	303.8
1955	276.3
1956	236.4
1957	216.2
1958	189.8
1959	148.2
1960	100.0
1961	84.4
1962	64.1
1963	53.3
1964	39.1
1965	31.9

A second valuable benefit provided by space and defense demands was experience with new types of computer applications. Military requirements for space, speed, and reliability resulted in more sophisticated applications than those first sought by business firms. Typically, the techniques used in a particular computer application would initially be developed and refined in a military system. Subsequently, the same techniques would be applied to a commercial data processing system. The most notable example of applying military computer experience to commercial problems is in the area of real-time computer systems. The first commercial real-time computer systems were descendants of the Air Force's pioneering real-time air defense computer system — the semiautomatic ground environment or SAGE.

New Industries: Semiconductors and Computers

An executive of a major computer company, with substantial experience in both the space-defense and the commercial markets, stressed in an interview with the author the relationship between applications in the two markets when he said: "Space and defense computer applications have served as a 'crystal ball' for predicting the future direction of computer use in industry." He based his statement on his premise that "because many computer applications in the space and defense programs are on the leading edge of the technology, they point the way to future applications in industry — and elsewhere in government — two to three years hence."

A third benefit of space and defense demands was the "demonstration effect." Sales in the commercial computer market depended on the acceptance of complex systems by business executives. Many of these executives lacked the technological knowledge needed to fully appreciate the potential of computers. The combination of unfamiliarity and skepticism among business executives would have been an immense barrier to commercial sales were it not for the "demonstration effect" of space and defense computer systems. In the 1950s military demonstrations of computer ability accelerated the business world's acceptance of the computer. More recently the widely publicized use of computers for NASA space missions during the 1960s produced an unparalleled demonstration effect. One computer industry observer, the computer engineering professor quoted earlier, noted the significance of NASA's "successful use of computers, particularly on Apollo [which] greatly increased industry's receptivity to computers." Moreover, he found that managers became "more willing to let computers control things, such as in process control of oil refineries," and added that "the momentum generated by Apollo had a big effect on confidence, producing a sudden turnaround in industry receptivity."

Competition and the Support of New Firms

Space and defense demands have not only stimulated the growth of the semiconductor and computer industries, they have also helped stimulate and maintain competition in these fields. The ability of new firms to enter the market is an essential characteristic of competitive industry. In both the computer and the semiconductor industries, government demand had been a factor in supporting new firms, but space-defense stimulation of competition is more pronounced in the semiconductor market.

The first commercial producers of semiconductors were Western Electric and such major tube manufacturers as General Electric, RCA, and Sylvania. General Electric, Raytheon, and RCA began producing transistors in 1951, and by 1954 eight different firms were manufacturing transistors. Other firms were slower to respond to the new technology, but they entered the market in large numbers during the mid-1950s. The three types of firms which entered the market were: (1) firms which were large and diversified before they entered the semiconductor industry, such as Hughes, Motorola, IBM; (2) small firms engaged in other branches of electronics or in other industries, such as Texas Instruments, General Instrument; and (3) new companies which were initially established to manufacture semiconductors, such as Transitron, National Semiconductor (Tilton, 1971, pp. 65-69).

Many of the new firms were able to enter the market because the military agencies were willing to buy from new and untried firms. By 1957 these new firms accounted for 64 percent of the total market and 69 percent of the government market (Table 2-5). The two largest firms in the market, Texas Instruments and Transitron with 20 and 11 percent market shares respectively, had both benefited greatly from military demand. In early 1953, the military authorized the use of Transitron's gold-bonded diode before the company had made any significant commercial sales (Harris, 1959, p. 142). In 1954, Texas Instruments, building on research at Bell Laboratories, introduced the first silicon transistor. The innovation opened a vast new market, particularly in products for military use.

Space program demand has been especially important in establishing a competitive market for metal oxide semiconductor (MOS) devices. The first MOS transistor was introduced by Fairchild in 1960. However, as a result of several technical difficulties and the departure of many MOS specialists, Fairchild dropped these devices from their product line after two years of commercial production. Then, beginning in 1963 the new firms quickly entered the field. General Micro Electronics was founded by Fairchild personnel in 1963 and received one of its first contracts to produce MOS integrated circuits from NASA. When

Table 2-5

U.S. Semiconductor Market Shares of Major Firms[a]

(Selected years)

Type and Name of Firm	Percentage of Market			
	1957	1960	1963	1966
Western Electric	5%	5%	7%	9%
Receiving tube firms:				
General Electric	9	8	8	8
RCA	6	7	5	7
Raytheon	5	4	(b)	(b)
Sylvania	4	3	(b)	(b)
Philco-Ford	3	6	4	3
Westinghouse	2	6	4	5
Others	2	1	4	3
Subtotal	31	35	25	26
New firms:				
Texas Instruments	20	20	18	17
Transitron	12	9	3	3
Hughes	11	5	(b)	(b)
Motorola	(b)	5	10	12
Fairchild	(b)	5	9	13
Thompson Ramo Wooldridge	(b)	(b)	4	(b)
General Instrument	(b)	(b)	(b)	4
Delco Radio	(b)	(b)	(b)	4
Others	21	16	24	12
Subtotal	64	60	68	65
TOTAL	100	100	100	100

[a] Market shares are based on company shipments and include in-house government sales.

[b] Not one of the top semiconductor firms for this year. Its market share is included in the "others" category.

Source: John E. Tilton, *International Diffusion of Technology: The Case of Semiconductors* (Washington, D.C.: The Brookings Institution, 1971), p. 66.

Philco-Ford acquired General Micro Electronics in 1966, several of the latter's employees left to form American Micro-Systems which was subsequently awarded a NASA contract. The firm developed the technology required to produce stable MOS devices and became one of the early suppliers of specialized commercial MOS devices. American Micro-Systems eventually became a major source of such devices throughout the country (Denver Research Institute, 1972, pp. 151-55).

The impact of space and defense demands on competition among computer firms is less pronounced than among semi-conductor producers, but the impact is evident, particularly in the industry's early years. Remington Rand, the company which delivered the first UNIVAC computer, had a head start on IBM during the early 1950s. IBM did not enter the electronic computer industry until 1953, but it was the clear industry leader by 1956. Most analyses attribute the overwhelming IBM success to the firm's aggressive salesmanship. While essentially correct, these analyses ignore the significance of space and defense demands for IBM. Space and defense purchases exceeded 30 percent of the firm's computer revenues in the mid-1950s and did not drop below 20 percent until 1958 (Burck, *et al.*, 1965; Bernstein, 1966, pp. 81-82; and Gropelli, 1970).

During the late 1950s space and defense sales were vital to the growth of many new computer firms. The most significant of those was the Control Data Corporation begun by former Sperry Rand employees in 1957 as a small but successful supplier of standardized systems for military and scientific use. Control Data Corporation raised revenues from almost nothing in 1958 to around $121 million in the fiscal year ending June 1964. By 1965 the firm had displaced RCA as the third largest computer manufacturer — behind IBM and Sperry Rand. Space-defense business provided the initial impetus for Control Data Corporation's growth; almost all of the company's pre-1960s sales were to government agencies (Burck, *et al.*, 1965, pp. 89-90; *Wall Street Journal*, 1963, p. 32; and *Electronic News*, March 6, 1967, p. 29).

During the 1960-70 decade, space and defense demands aided several computer manufacturers, particularly Sperry Rand and Control Data Corporation, to maintain or improve their competitive position vis-à-vis IBM. A 1962 government policy shift from computer rental to the direct purchase of computers promoted the growth of IBM's competitors. In 1963 IBM began to feel the impact of this policy when several space and defense agencies converted from IBM to other equipment. In 1961, IBM was responsible for 81 percent of all computers supplied to the government. By 1966, IBM's net shipments to government agencies were negative as a result of returned rental equipment which was replaced by new equipment purchased from IBM's competitors ("Something to Think About," 1963; and Gropelli, 1970, p. 215).

The market share data demonstrate the momentousness of space and defense sales for IBM's competitors. Approximately a fourth of all installed UNIVAC computers were located in space or military agencies by 1970. The comparable percentages for the other manufacturers were 23 percent for Burroughs, 16 percent for Control Data Corporation, 15 percent for RCA, and 7 percent for IBM. New companies, such as Scientific Data Systems (subsequently Xerox Data Systems) benefited substantially from space and defense demands; in 1971, 28 percent of all Scientific Data Systems' computers were located in space and defense agencies. Other smaller companies such as EMR Telemetry, Electronic Associates, and Autonetics all had market shares which exceeded 15 per cent (Gropelli, 1970, p. 78; and U.S. Congress, House, 1965 and subsequent annual editions). Table 2-6 shows the computer stock value of space and defense agencies as a percentage of the total computer stock value. All market data are based on the value of installed equipment at the end of each year.

Technological Impacts

Although economic impacts are an important part of government influence on private industry, they do not tell the entire story. Essentially, they provide information regarding only the flow of research funds and the distribution of sales revenue. Yet government programs can also provide important technological support. In the case of semiconductor and computer industries, the technological benefits for individual firms were significant.

It is important to note that for semiconductor manufacturers, learning by doing on the part of production workers is a relatively unimportant source of learning economies. Most semiconductor

Table 2-6

Space-Defense Computer Stock Value as a Percentage
of Total Computer Stock Value [a]

(Selected years)

Firm	1960	1961	1963	1964	1965	1968	1970	1971
				Year				
IBM	10	10	8	8	8	7	7	5
Burroughs	18	19	8	5	14	10	23	18
Control Data Corporation	5	6	10	13	20	14	16	14
General Electric	0	0	5	4	8	3	5	0
Honeywell	0	2	10	7	9	8	5	4
RCA	15	14	13	15	17	12	15	15
Univac	6	10	11	10	13	15	24	21
Scientific Data Systems	—	—	11	8	4	10	21	28
EMR Telemetry	—	—	18	9	14	5	24	27
Electronic Association	—	—	—	—	0	12	14	15
Autonetics	13	8	13	22	20	26	15	15
Digital Equipment	—	—	9	9	3	2	2	3

[a] Value of computer stock computed by using annual inventory data on number of units installed for each computer model and multiplying this figure by estimated monthly rental value for that model.

Source: A. Gropelli, "The Growth Process in the Computer Industry" (New York: New York University, 1970), doctoral dissertation; and U.S. House of Representatives, 89th Cong., 1st Sess., *Inventory of Automatic Data Processing Equipment in the Federal Government* (Washington, D.C.: Government Printing Office, June 1965).

learning economies are not an automatic by-product of production but, rather, are produced by the deliberate efforts of firms to upgrade production technology. Such economies are generated by improvements in the production process arising from experimentation with the length and temperature of the crystal preparation process, the measures taken to control crystal contamination, the equipment used in assembly and testing, the scheduling and organization of operations, and so on.

The defense and space programs demanded high-quality components from the standpoint of both reliability and performance. The challenging specifications set by the military agencies and NASA accelerated many of the learning economies related to improved production processes. Military and space programs established clear specifications for the types of components they required. This prompted many companies, including newer ones, to develop semiconductors which were at least as good as, if not better than, those already under government contract (Tilton, 1971, pp. 85-87).

Whereas military demand accelerated the advance of second-generation component technology, the U.S. space program was dominant in accelerating the development of third-generation technology. The development of the Apollo Guidance Computer, a vehicular-borne computer, required extensive miniaturization and extremely high reliability. The M.I.T. Instrumentation Laboratories served as NASA's prime contractor for this project. Twenty-two different component manufacturers participated in the procurement process for the Apollo Guidance Computer. Several of these firms worked with M.I.T.'s Instrumentation Laboratories for seven or eight years, although their components were never purchased by NASA. The firms remained potential suppliers for such a long period of time because of the learning experience generated by NASA's demanding technical specifications. While NASA's approach to component reliability did not alter commercial production processes in the short run, over the longer run firms did eventually redesign production processes and install new equipment in order to transfer space-defense reliability methods to commercial production.

NASA's efforts to improve the reliability of MOS devices illustrate the technological influence of space program demand on the semiconductor industry. The NASA Microelectronics Reliability Program was instituted to set criteria for the acceptance of semiconductor devices and required NASA inspection of the production facilities of semiconductor suppliers. All firms interested in supplying MOS devices to NASA had to comply with the NASA Microelectronic Reliability Program specifications. If a company was unable to meet the acceptance criteria, NASA aided it in making the necessary changes to meet reliability standards. Major integrated circuit suppliers, such as Motorola, Fairchild, and Harris, participated in the program and used NASA's technical assistance to improve their processing capabilities (Denver Research Institute, 1972, pp. 153-55; and interviews with executives of NASA, M.I.T.'s Instrumentation Laboratories, and semiconductor firms).

Space and defense demands also influenced technological developments in the computer industry. Work on military and NASA projects accelerated developments in two notable areas of computer technology: (1) real-time operating systems and (2) computer system reliability.

The first — and still most famous — real-time computer system was the Air Force's SAGE. The system was set up to protect the United States against a surprise air attack. The SAGE system consisted of a network of radar-fed computers that continuously analyzed every cubic foot of air space around the United States, instantly tracked all approaching aircraft, and decided on an appropriate response. After seven years of development work by M.I.T.'s Lincoln Laboratories, with the assistance of IBM and Burroughs, the first of sixteen SAGE centers was completed in 1958. The original computer programs for SAGE consumed eighteen hundred man-years of effort; the total system development cost has been placed at nearly $2 million.

There is virtual unanimity throughout the data processing community that all subsequent real-time computer systems owe a great deal to SAGE. American Airlines implemented SABRE, its $30 million real-time seat reservation system, in 1964. It is unlikely that SABRE could have been developed at that early date without the experience provided by SAGE.

In the mid-1960s space applications enabled computer manufacturers to further expand their real-time capabilities. Most of the early computer operating systems, which are the set of programs provided to the user by the computer manufacturer for purposes of operating the system, were in a "batch" mode as opposed to a "real-time" mode. For both IBM and Control Data

Corporation, it was NASA's demanding specifications that forced them to advance their capabilities in providing "real time" operating systems. In fact, IBM's first experience in developing a real-time operating system was with NASA, and the technology developed for that system soon became the industry standard. Control Data Corporation's first real-time operating system was also for NASA, at the agency's Langley Research Center. The Langley center is Control Data Corporation's largest single installation, and several features of NASA's Langley system have become an integral part of the corporation's standardized real-time operating system (interviews with executives at IBM, Control Data Corporation, Univac, and several NASA computer centers).

Improved reliability of computer operations is a second area where technological gains were accelerated by space and defense demands. At IBM, Control Data Corporation, and Univac, NASA's technical specifications prompted significant improvements in the reliability of large computer systems. Indeed, IBM's major involvement in the space program has been with the computer complex at NASA's Manned Spacecraft Center in Houston. This system collects, processes, and sends to Mission Control the information needed to direct manned space missions, such as Apollo and Skylab. The initial reliability specification for this computer system was 41.7 hours of operation without a major breakdown.

Originally the IBM system was unable to meet this specification. Special teams of IBM production and test personnel were assigned to work with NASA reliability specialists to correct the problems. By paying extraordinary attention to every aspect of reliability, these teams raised the reliability of the real-time computer complex to a hundred hours of operation without a catastrophic breakdown — a level which was two and a half times the initial specifications! These new reliability features prompted and financed by NASA, were quickly incorporated into IBM's commercial product line.

Manpower Impacts

Although economic and technological impacts have been the dominant concern in most efforts to assess the influence of government programs on industry, they do not fully cover the range of possible impacts. Public programs also influence the supply of manpower in industrial firms. These manpower impacts occur in one or more of three ways:

(1) *Direct supply:* A government program can be a direct source of experienced manpower for companies in a particular industry. Personnel may receive training and operating experience in a government program and eventually leave to join a private firm.

(2) *Interfirm mobility:* If government demand enables new firms to enter an industry, a high rate of interfirm mobility may result as personnel leave established older firms to join newer firms.

(3) *Intrafirm mobility*: A firm's work on a government program may provide advanced training for its employees. If the firm seeks to apply the expertise of these employees to other government or commercial projects, a high rate of intrafirm mobility may result.

In the electronics industry the manpower impacts of space and defense demands have taken two of these forms. In semiconductors, the willingness of space-defense procurement officials to purchase semiconductor devices from small new firms has facilitated high *interfirm mobility*. In contrast, the influence of space and defense programs on computer industry manpower has been to stimulate *intrafirm mobility*. Large space and defense computer systems have provided an important training ground for project managers and engineers who later applied their knowledge and experience to other large government and commercial computer development projects which were awarded to their firm.

Interfirm mobility in the semiconductor industry includes the movement of personnel among existing firms as well as the departure of engineers and managers from established firms to set up their own companies. Bell Laboratories, Hughes, Philco, RCA, Sylvania, Texas Instruments, and Motorola have all experienced the defection of key employees who set up new firms. Fifteen separate new firms, which were formed between 1952 and 1967, can be traced back to Bell Laboratories. The first new semiconductor firms formed by former Bell personnel were Transitron in 1952 and Shockley Laboratories in 1955. Fairchild Semiconductor, which spun off from Shockley

Laboratories in 1955, had five direct and four indirect spinoffs over the next twelve years (Figure 2.l).

Mobility among existing semiconductor firms dates back to the movement of Bell Laboratories employees to Texas Instruments and Sylvania during the early days of the industry's development. During the 1960s, Fairchild was the focal point of much industry mobility. A number of Fairchild executives departed in 1967 to revitalize the National Semiconductor Corporation, which had fallen on hard times. A year later Fairchild recruited the general manager of Motorola's semiconductor division and a number of other Motorola executives (Hoefler, 1968; "Fight That Fairchild Won," 1968, p. 106; and Tilton, 1971, pp. 77-81).

Bell Laboratories has played an important role in the semiconductor industry's high rate of interfirm mobility. As the research arm of AT&T, Bell Laboratories has contributed a disproportionately large share of the major semiconductor product and process innovations. Yet AT&T did not use its strong patent position to prevent the entry of new firms into the market; instead, the company adopted a liberal licensing policy and expedited the transfer of technology in a number of ways. It also established the precedent for the industry's tolerant attitude regarding the departure of scientists and engineers to either found new firms or accept positions in competing firms. (Tilton, 1971, pp. 73-81, suggests that the government's antitrust suit against AT&T was a major factor involved in the company's decision to rapidly disseminate its new semiconductor technology. As part of the 1956 Consent decree that culminated lengthy antitrust litigation, AT&T surrendered the right to sell semiconductors in the space and defense markets and to produce for its own needs.)

In the computer industry there is substantial intrafirm mobility in the form of project managers and engineers who move from space-defense divisions to the commercial side of the business. Alumni of IBM's NASA real-time computer complex have gone on to direct the development of major computer systems for the Federal Aviation Administration, the Bank of England, and New York City. At Univac, approximately two hundred people per year are employed full time on NASA work. Each year a number of these employees move to another area of

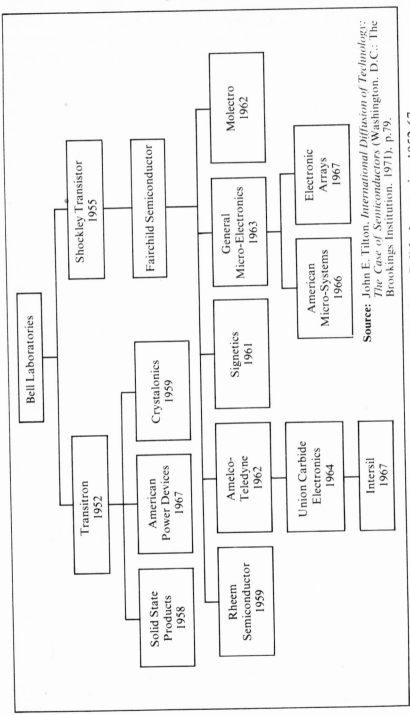

Figure 2.1. Semiconductor Firms Descending from Bell Laboratories, 1952-67

Source: John E. Tilton. *International Diffusion of Technology: The Case of Semiconductors* (Washington. D.C.: The Brookings Institution. 1971). p.79.

the company; consequently Univac estimates that several thousand of its current employees have previously worked on some aspect of NASA systems. One specific example is the project manager of a large computer network developed by Univac for Bell Telephone who previously served as the firm's project manager for NASA's Goddard Space Flight Center. At Control Data Corporation, several engineers who acquired real-time simulation experience at NASA's Langley Research Center went on to develop similar applications at Volkswagen and Grumman Aircraft (interviews with executives of the various firms previously mentioned).

Summary

Large government programs in defense and space have greatly aided U.S. firms in their rise to a dominant position in the world electronics industry. The most important impact of these federal programs was the funding of research and development and the initial demand they provided for computers and semiconductors during the start-up years of both industries.

Direct and indirect financial support for research and development by space and defense programs was an important influence in the development of the semiconductor and computer industries. Direct financial support for semiconductor research and development totaled $66 million between 1955 and 1961. These government grants helped semiconductor firms to greatly expand their production capacity during this crucial six-year period. In addition to this direct research and development funding, semiconductor firms received indirect federal research and development support through subcontracting by weapons systems prime contractors. The Defense Department estimates that the research and development subcontracts awarded by its prime contractors were more than equal to the direct research and development expenditures. At the end of the 1950s the total of direct and indirect government-financed research and development was approximately a quarter of total semiconductor industry expenditures for research and development.

Federal agencies, particularly the military services, provided strong financial support for every major U.S. computer development between 1945 and 1955. The Army supported the development of ENIAC, the first electronic computer, for use in trajectory calculations. The major technological advances during the first ten years of electronic computers were achieved as part of the effort to develop large computers which met the specifications set by military and other government agencies. Most of these advances were subsequently incorporated into the medium and small-scale computers designed for the commercial market. The large U.S. government outlays for computer development during this period dwarf those of other countries, such as Great Britain, and help account for the early dominance of U.S. firms in the computer industry.

The second way in which federal space and defense programs influenced the computer and semiconductor industries was by generating huge markets for these products. Space and defense demands were a major factor in the growth of the U.S. semiconductor industry, an industry where learning economies proved essential. Learning economies have resulted in dramatic decreases of semiconductor prices; the average price of an integrated circuit dropped from $50 in 1962 to $0.63 in 1973. During the early years of second- and third-generation component technology, the space and defense markets provided a substantial part of the sales volume which made these learning economies possible. Space and defense demands accounted for at least 35 percent, and as much as 45 percent, of semiconductor sales for each year between 1955 and 1961, and over 70 percent of annual sales during the first four years of integrated circuit production.

The market for military data processing systems had reached the $200 million level before Remington Rand delivered the first UNIVAC computer for business data processing in 1954. The space and defense markets accounted for more than 60 percent of all computer sales during the industry's first decade, and the sales of commercial computers did not overtake space-defense hardware sales until 1962.

As both the computer and semiconductor industries matured, space and defense demands promoted competition among existing firms and aided the entry of new firms. The receptivity of military agencies and NASA to the products of new semiconductor companies enabled these companies to enter the market easily. Several new companies used space and defense contracts to establish their initial market position. The first sales for Texas Instruments' silicon transistor and for Transitron's gold-bonded diodes were for use in military products.

Control Data Corporation, a firm which was the third largest computer manufacturer by 1965, depended exclusively on military sales when it first entered the industry in 1957. Space-defense business helped IBM's major competitors — particularly Univac, Control Data Corporation, and Burroughs — to improve their market position during the 1960s. By 1970, a fourth of all UNIVAC computers were in either space or defense installations.

In addition to generating needed sales revenues for firms during the early stages of growth, space and defense demands accelerated the advance of semiconductor and computer technology. The learning economies, which have been so significant in the semiconductor industry, have not been an automatic by-product of production. They have required deliberate planning. The challenging performance and reliability specifications set by the military agencies and NASA accelerated many of these semiconductor learning economies. The space program's specifications for the integrated circuitry of the Apollo guidance computer was a major impetus for improvements in the reliability of third-generation component technology.

Space and defense demands also accelerated developments in two other valuable areas of computer technology — real-time operating systems and computer system reliability. The techniques used in the earliest commercial real-time systems were derived from SAGE, the Air Force's pioneering computer system. More recently, the demanding specifications of NASA's computer system caused both IBM and Control Data Corporation to improve their real-time operating system capabilities. A prominent example of reliability improvements is the computer complex which IBM installed at NASA's Manned Spacecraft Center in Houston. Under the pressure of NASA requirements, IBM improved system reliability two and a half times initial levels; these reliability improvements were subsequently incorporated throughout IBM's 360 product line.

Finally, space and defense demands have facilitated the development and mobility of professional manpower in the computer and semiconductor industries. The willingness of space-defense procurement officials to purchase semiconductor devices from small new firms often stimulated engineers and managers from established companies to set up their own firms. The number of new semiconductor firms which emerged was very large; Bell Laboratories alone was the direct and indirect source for fifteen new firms formed between 1952 and 1967. The impact of space and defense programs on computer industry manpower has been to stimulate intrafirm as opposed to interfirm mobility. In the major computer firms, space and defense projects served as an important training ground for project engineers and managers who were later able to transfer these skills to commercial projects.

The significance of space and defense demands for the U.S. electronics industry can only be perceived by reviewing each type of impact — economic, technological, and manpower. The government's early research and development support, its assured demand during start-up periods, its encouragement of competition, its acceleration of technological progress, and its facilitating professional mobility are the separate impacts which together helped place U.S. firms in a position of world leadership in the electronics industry.

References

Arthur D. Little, Inc. *Patterns and Problems of Technical Innovation in American Industry.* Prepared for the National Science Foundation. Washington, D.C.: Arthur D. Little, Inc. 1963.

Bernstein, J. *The Analytical Engine.* New York: Vintage Books, Inc. 1966.

Burck, Gilbert; *et al. The Computer Age and Its Potential for Management.* New York: Harper & Row, Publishers, Incorporated. 1965.

Chow, Gregory C. "Technological Change and the Demand for Computers." *American Economic Review* (December 1967).

Denver Research Institute. *Mission-Oriented R&D and the Advancement of Technology: The Impact of NASA Contributions.* Volume II. Denver, Colorado: Denver Research Institute. May 1972.

Electronic Industries Association. *Electronic Industries Yearbook, 1969.* Washington, D.C.: Electronic Industries Association. 1969.

Electronic News (March 6, 1967).

"Fight That Fairchild Won." *Business Week* (October 5, 1968).

Freeman, C. "Research and Development in Electronics Capital Goods." *National Institute Economic Review* (November 1965), vol. 34, no. 40.

Gropelli, A. "The Growth Process in the Computer Industry." New York: New York University. 1970. Doctoral dissertation.

Harris, William B. "The Company That Started with a Gold Whisker." *Fortune* (August 1959).

Hoefler, Don C. "Semiconductor Family Tree." *Electronic News* (July 8, 1968).

Jacobowitz, H. *Electronic Computers*. New York: Doubleday & Co., Inc. 1963.

MacLaurin, W. R. *Invention and Innovation in the Radio Industry*. New York: The Macmillan Company. 1949.

Postan, M.; Hay, D.; and Scott, J. "Design and Development of Weapons." In *History of the Second World War*. Part III, Chapter XV. London: Her Majesty's Stationery Office. 1964.

"Something to Think About." *Barrons* (July 8, 1963).

Texas Instruments, Inc. "First Quarter and Stockholders Meeting Report." Special edition. April 18, 1973.

Tilton, John E. *International Diffusion of Technology: The Case of Semiconductors*. Washington, D.C.: The Brookings Institution. 1971.

U.S. Congress, House. *Inventory of Automatic Data Processing Equipment in the Federal Government*. Prepared by the Bureau of the Budget for the Subcommittee on Census and Government Statistics. 89th Cong., 1st Sess. Washington, D.C.: U.S. Government Printing Office. June 1965.

_____, Senate. *Patent Practices of the Department of Defense*. Preliminary report of the Subcommittee on Patents, Trademarks, and Copyrights. 87th Cong., 1st Sess. Washington, D.C.: U.S. Government Printing Office. 1961.

U.S. Department of Commerce, Business and Defense Services Administration. *Electronic Components: Production and Related Data, 1952-1959*. Washington, D.C.: U.S. Government Printing Office. 1960.

Wall Street Journal, January 17, 1963.

Chapter Three

Transformation of a Science: NASA's Impact on Astronomy

In the years since the end of World War II, astronomy has experienced a period of discovery unmatched since Galileo first turned his simple astronomical telescope toward the sky. During its short lifetime, NASA has contributed the means by which astronomers have made some notable discoveries, and the agency's continuing contribution promises more advances for astronomy in the future. Equally as meaningful as its contributions to the development of the science have been NASA's effects upon the structure and organization of the astronomical community.

In the 1950s, radio astronomers began to reveal a startling new and different universe from the one seen by visible light. With their new ground-based instruments, they discovered radio galaxies more than a million times brighter than our own; they found the faint, attenuated radiation of the great fireball from which the universe was born; they traced the grand spiral pattern of the Milky Way, obscured to optical telescopes by dust; they identified clouds of complex interstellar molecules. To their amazement, they found mysterious quasars (celestial bodies) that may be far enough away to be galaxies but behave like stars — and the strangeness of quasars is matched by pulsars, believed to be small, super-dense concentrations of matter, as massive as the sun but only miles in diameter, spinning from one to thirty times a second.

81

Even as these discoveries were changing our view of the universe, space observations of high-energy radiation revealed more unexpected phenomena. Since the mid-1960s, space instruments have allowed astronomers to observe explosive, hot events in a universe of cosmic violence and exploding galaxies, filled with omnipresent high-energy particles and twisting magnetic fields. The observations even suggest such peculiarities as "black holes," stars in a state of relativistic collapse from which no light or radiation of any kind can escape.

With the entire electromagnetic spectrum opened to observation over the past decade, astronomy has entered what has been widely recognized in the science as a new golden age. The National Science Board assured Congress in its 1970 report that "the rapid pace of discovery in astronomy and astrophysics during the last few years has given this field an excitement unsurpassed in any other area of the physical sciences" (National Academy of Sciences, vol. 1, 1972, p. 55).

Scientists realize that NASA has contributed to the exciting discoveries in astronomy; NASA's financing has provided the resources that have made the golden age more than just a metaphorical term. However, NASA was not the initiating agency in radio astronomy, and its support of radio observations has been modest. Not until the 1960s, when rockets and satellites allowed observations from above the earth's atmosphere, did NASA add its own large and unique contributions to astronomical discoveries. Its technology affected astronomy far more than did that of radio observations, for NASA's contributions were much more extensive and complex. In developing that technology, NASA began to transform the science through the vast and varied demands it had to make upon a small scientific community; the resources it made available to meet these demands loomed immense beside those that astronomers previously had available.

Through NASA's programs scientists were encouraged to resurrect moribund subfields of astronomy grown old from lack of new data. With NASA support scientists were able to discover new, previously unknown phenomena that required new kinds of expertise and analysis; NASA's comprehensive and widespread encouragement of space-related sciences enhanced the quality and productivity of active subfields. The demands of space technologies changed the very nature of large parts of astronomy, fractionating it into well-defined specialties even as it integrated other specialties into new groupings. Most strikingly it created — or is creating — out of a small, university-bound science, a Big Science, with all of the problems of manpower, organization, and public policy that such a development entails.

Astronomy before NASA

Until recently astronomers have numbered their membership in the hundreds; not until the late 1960s could they count themselves by the thousands (National Academy of Sciences, vol. 2, 1972, pp. 334-35). Thus astronomy has indeed been a small, slowly growing science. One indication of that growth is the 4.5 percent annual increase of American membership in the International Astronomical Union for the years 1923 to 1955; after 1955 into the 1960s growth jumped to 8 percent a year. Another indicator of both the small size and slow growth of the astronomical community in the past is the number of doctorates awarded annually in astronomy. In the 1920s universities produced fewer than ten a year, increasing through the 1950s at a rate of only 3.8 percent a year. The total number of doctorates (in astronomy) awarded annually was smaller than those in physics, with the growth rate hardly more than half as large (National Academy of Sciences and National Research Council, 1964, pp. 30, 31).

The number of astronomers was small because there were not sufficient funds to support them, and few facilities with large telescopes were available for advanced observation and study. Until 1950 Americans had built only thirteen optical telescopes with apertures equal to or larger than 36 inches. Six of these had been built before 1920, three of them before the turn of the century! In the 1950s another six were completed. The opportunity for the science to absorb a sizable number of optical astronomers did not exist until the 1960s when 28 new instruments (36 or more inches) were built, increasing the number by almost one and a half times. A dozen more instruments of comparable size were to be installed in 1975 (National Academy of Sciences, vol. 2, 1972, pp. 388-93, in which the authors warn that the list was only "reasonably accurate as of September 1971; but since additions and modifications occur frequently, some inaccuracies are bound to be present").

Before the mid-1950s only a few astronomers had any but limited access to large instruments. The three or four institutions that controlled the big telescopes, sixty inches or larger, dominated the field in every way until recently. As a consequence, astronomers were described by an informed observer (in Metz, 1973, p. 247) as

> . . . a small number of men [and women] largely isolated from the rest of science, who were supported for the most part by private funds and were comfortable with projects that spanned decades. The profession has always been a consortium of colorful individuals, a community with a tradition of sharp, sustained, and at times personal debates. . . . [W]ith typical measured pace, the deans of astronomy discarded their 19th-century style of doing science.

Astronomy in the nineteenth century and in the first half of the twentieth century remained a purely observational science with few observing instruments. Astronomers did not design experiments as physicists might; nor did they manipulate samples as chemists do. They were seldom encouraged to consider improvements in instruments; first, funds were too limited to conduct such work, and second, most astronomers enjoyed only limited access in working with a major telescope. Furthermore, the number of astronomers was so small and the problems so many that these persons could specialize in an area of research with little likelihood that another would encroach upon or compete with them. They could easily keep busy and productive using the available, familiar techniques and the instruments provided them. They were not pushed to seek new approaches to their research, and their penury persuaded them that their wisest course was to exploit to the fullest the tools at hand. As late as 1964 astronomers complained that so few major optical telescopes were available "that no more than two or three astronomers in the entire world now have the opportunity to work on the most exciting problems in a given field" (National Academy of Sciences and National Research Council, 1964, p. 16). Such circumstances kept astronomy a small science and almost guaranteed that it would lag in adapting technology developed elsewhere.

An example of American astronomers' lagging response to technological advance is the case of radio astronomy. Until the 1940s, astronomers contented themselves with a technology that was limited to photographic, simple photoelectric, bolometric, and low- or high-resolution spectroscopic measures of photons in one octave of the electromagnetic spectra, the optical "window" in the earth's atmosphere. Meanwhile electronic discoveries and advances had made new kinds of detectors possible, but only for those with adequate funding and acquaintance in the new technologies.

The panel on radio astronomy in the 1973 Greenstein report (National Academy of Sciences, 1972, vol. 2, p. 22) asserts that the science of radio observations "had its origins in the universities"; the originators were not astronomers, however. University astronomers became interested in the new techniques only after electrical engineers and physicists began to explore radio astronomy's potential. In addition, radio observations became a part of astronomy as scientists from other fields moved into astronomy and students enlarged their training to include study of the radio branch of science.

Karl G. Jansky, of the Bell Telephone Laboratories, in 1931 first discovered — with primitive instruments — that radio waves from extraterrestrial sources could be detected. In the late 1930s a radio amateur, Grote Reber, constructed a 32-foot paraboloid that he set up in his back yard in Wheaton, Illinois; with it he was able to trace a rough map of the sky's radio contours. By the end of World War II, radio technology had developed to a sophisticated level, and many able people had become familiar with it. Talented European and Australian scientists were particularly innovative in adapting the wartime electronic developments for use in radio astronomy. They soon discovered discrete radio sources and other phenomena that encouraged them to build large radio telescopes and antenna arrays.

In the early 1950s, the Naval Research Laboratory began a modest radio astronomy project, building the first fifty-foot parabolic antenna. Cornell and Harvard universities soon followed. In 1951, Harvard researchers detected a 21-centimeter radiation of hydrogen, and in 1953 the university graduated its first doctoral student in radio astronomy (National Academy of Sciences and National Research Council, 1964, p. 5).

The major impetus for the new branch of astronomy, certainly for the early experimentation and development, came from outside the small, well-established core of scientists who made up the astronomical community. An eminent radio astronomer, whose training was in electrical engineering, explained in an interview with the author: "We knew the technology and wanted to see what could be done with it." But astronomers, he found, "looked down on it. They were not particularly familiar with it"; and because at first it did not appear that it would tell them more "about stellar activity or the galaxies than they already knew, they weren't interested in developing it." And he added, "They probably weren't capable of developing it in any case. The concepts [i.e., the electronic technologies] were foreign to them."

The radio astronomers were scientists who knew about the instruments and understood engineering. They worked closely with electronics experts in government and industry who were pushing the development of the new technology. Much more so than the optical astronomers, radio astronomers were a part of the technological complex that developed in the 1950s. By the end of that decade, the government turned to them for help. When instruments were needed to track the first Russian Sputniks, the "dishes" and equipment of the radio astronomers could be — and were — copied.

The development of radio astronomy needed more than trained, able manpower and a promising technology — it needed funding. The success of the radio astronomers depended upon the willingness of the Defense Department, particularly the Navy and Air Force, in the 1950s to support the expensive installations.

The Armed Services also invested considerable sums in university-managed and -manned optical installations, making possible research activities and the acquisition of equipment and buildings that astronomy departments had never before been able to consider. An example is a contract granted by the Air Force in 1952 to the Yerkes Observatory for studies of upper atmosphere phenomena such as air glow and auroras. Although the dollar amount of the contract was less than half a million dollars, it was larger than all the rest of the observatory's budget combined. With the appearance of such resources, astronomers could begin investing in new and advanced equipment and take advantage of new technologies.

Through the early and mid-1950s, a variety of government agencies — the Office of Naval Research, the Smithsonian Astrophysical Observatory, the Air Force, and the National

Science Foundation — began to fund astronomical research and new ground-based instruments, both optical and radio, on a scale that was sure to create profound changes in the science. The most significant investment was in the Kitt Peak National Observatory, a large public facility with six major optical telescopes, the largest of which is 158 inches, completed in 1973 (National Academy of Sciences, vol. 2, 1972, pp. 388-93, with the same caution as that expressed earlier by these authors).

Government funding for "space" experiments was limited throughout the 1950s. Those carried out had only small rockets or high-altitude balloons to carry experimental packages to the edge of the atmosphere. The Navy and the Air Force were the largest supporters of the space research that was conducted. The number of interested scientists was small. A few were persuaded that astronomers needed to consider further ways and means of observing from space. As early as 1940, Dr. Lyman Spitzer of Princeton pointed out the advantages of an optical telescope in space; and in 1957 Dr. Lloyd Berkner, a geophysicist, called upon astronomers to respond with experiments for space and suggestions on how the science should proceed to use space.

NASA's Missions and Their Effects on Astronomy

Many astronomers, like many scientists, were skeptical that NASA had much to offer them. After Sputnik, the Air Force had urged that the United States develop a manned satellite, and the National Advisory Committee on Aeronautics (the agency to which NASA was successor in 1958) had gone to work on its own designs for a manned space capsule. In the public and congressional debate preceding enactment of the Aeronautics and Space Act of 1958, the merits of some kind of manned space program won official approval. President Eisenhower assigned specific responsibility for developing and carrying out a manned venture into space to NASA. It proceeded with Project Mercury.

The Space Act required that NASA contribute materially to one or more objectives, the first of which was "the expansion of human knowledge of phenomena in the atmosphere and space." The third objective, "the development and operation of vehicles capable of carrying instruments, equipment, supplies, and living organisms through space," seemed likely to eclipse the first. The large engineering effort required for manned flight and the massive technology that had to be mobilized for it distressed many scientists. Such leading scientific spokesmen as Vannevar Bush, James R. Killian, and George B. Kistiakowsky had made clear their opposition to manned space flight. Its special, complex demands and enormous costs, they argued, would of necessity push other scientific interests and purposes aside. They doubted that manned space flight would gain enough additional advantages over unmanned space flight to justify its far greater expense.

Jerome B. Wiesner, who became special assistant for science and technology to President Kennedy, also opposed the manned space program. In a report to the President-elect in early 1961, he and the advisory committee which he headed recommended a lower priority than had been given by the outgoing administration to Project Mercury. It remains unclear whether Wiesner and his committee had enough faith that "space" is a realm in which many scientific achievements could be expected. Nevertheless, they recommended that it be presented to the public only for its "cultural, public service, and military importance" (Swenson, Grimwood, and Alexander, 1966, p. 305). Unmentioned were any scientific benefits that might recommend it to taxpayers and voters.

The President did not follow the recommendations of his science advisers. First, shortly after his inauguration, he chose James E. Webb (an experienced government administrator and business executive, but not a scientist) as the new administrator of NASA. Some critics argued that Webb lacked the technical background that would attract scientists and qualified engineers to work for NASA. The critics were wrong; but Webb's appointment made clear that NASA's mission was not just scientific explorations. Second, in May 1961, President Kennedy directed the attention of NASA, and the nation, toward a great technological feat — to land a man on the moon within a decade and return him safely to earth. As Swenson, Grimwood, and Alexander (1966, p. 363) comment, "So impressive and dramatic an enterprise was Apollo, so full of engineering and gadgetry, that the project seemed made to order for a new American destiny." It was not the kind of project that was apt to reassure many astronomers that they had a part to play and a place to fill in it.

Not a few scientists, astronomers included, were concerned about whether NASA could be a reliable supporter of science when committed to so huge an engineering task. They feared that the production demands and technical considerations would come first, whipsawing the scientists, shortchanging their interests, and neglecting their needs in times of crisis. Their view was well expressed by Dr. Gordon J. MacDonald in testimony before the Senate's Committee on Aeronautical and Space Sciences (U.S. Congress, Senate, 1968, p. 752): "The size and scope of the U.S. space program was not determined by scientific requirements but by the implicit recognition that a nation's power and vitality are measured by its technological achievements."

Webb never forgot that Congress had charged NASA with the responsibility to advance space science. He worked hard to persuade scientists, and particularly astronomers, to help him and NASA's staff to develop a scientific program that lay within the jurisdiction that Congress had assigned to NASA; he wanted a program that would be of value to them and also to the nation. At a 1967 meeting, he reminded astronomers who still expressed doubts about NASA's intentions that he could not give them everything or anything they wanted. Congress had several objectives to pursue through NASA, he said, not simply the enchantment of science.

"While we are here to push science," he told the astronomers, "we are not limited to science. We want this new [space] technology to be available to any and all who can use it" (from author's research notes). He reminded them that they must consider the second- and third-order benefits as well as the first-order benefits "in all our missions." And he expressed his concern about the role of astronomers by adding: "I have never thought of . . . [the astronomical experiments] as a piggyback. But . . . [they are] in a way piggyback to the United States' uses and purposes of the whole space program. The astronomical community is a piggyback mission in our use of these vehicles for the benefit of the whole country."

Skeptical that NASA's managers were truly interested in astronomy, doubtful that the gigantic space agency would serve the science, and seeing valuable but marginal benefits from space observations, the astronomical community approached the opportunities for space exploration less than enthusiastically. In a major report to the National Academy of Sciences and the National Research Council in 1964, the panel on Astronomical Facilities, with Dr. A. E. Whitford as chairman, considered only the need for ground-based instruments and research. While the panel recognized that the high-energy radiations only recently observed in space raised fundamental questions, it assured its audience that "progress will be made [in providing answers] by clever and aggressive use of telescopes of the largest size, equipped with detectors such as radio receivers, spectrographs, photometers, and photographic plates" (National Academy of Science and National Research Council, 1964, p. 3). Space observations, clearly, were to play a secondary, though complementary role, to the optical telescope. Space astronomy would contribute only "certain key data, inaccessible from the earth." Plainly, the report was greatly influenced by the older core of astronomers familiar with and comfortable in using optical telescopes and photographic plates for observing. Not until the Greenstein report of 1972 did the astronomical community give full recognition to the unique role observations in space can play in advancing the science.

Resurrection of Old Astronomical Fields

Those who most easily and readily identify themselves as astronomers are scientists who specialize in stellar phenomena. The early space experiments did not contribute much to their research, and probable future experiments did not appear to promise more than a useful complement to data gathered through ground-based instruments. Thus many of these prestigious astronomers had some reason to believe that the costs of involving themselves in space projects might be greater than the benefits. Other astronomers, in lower status astronomical fields, discovered that space experimentation and space projects offered them immediate and enticing rewards. Those in two very small, almost moribund fields — celestial mechanics and geodesy (the study of the size and shape of the earth) — were among the first to benefit from space explorations. Any planned useful activity in space required of those involved a thorough knowledge of the physical characteristics of the environment. The shape, size, and curvature of the earth had to be known more accurately than in earlier times, for they affected orbits of satellites; success in monitoring, and certainly in guiding, a satellite or recovering it required an ability to predict its location at all times with a very narrow range of error. To fulfill most of its objectives in space, the space agency needed more and better services from experts in geodesy and celestial mechanics than had ever before been rendered.

The second American satellite, Vanguard I, launched in March 1959, proved the usefulness of space experiments to geodesists. Study of its orbits indicated that the earth is slightly pear-shaped, bulging in the aqueous southern hemisphere to a degree unsuspected before. Through triangulation measurements of satellites' location from earth, one could refine and check surface surveys; by measuring perturbations of satellite orbits geodesists were able to map the zonal harmonics of the earth's gravitational field and thus reveal its bulges and other anomalies. Geophysicists also learned much about the upper atmosphere previously unknown; for example, atmospheric density proved greater, by a factor of 5, than previously assumed, and the atmosphere was found to bulge on the side of the earth facing the sun. In general there exists a correlation between solar activity and atmospheric density at high altitudes.

The military services were exceedingly interested in the advances of geodesy, since accurate maps are essential in the effective use of many weapons. The Air Force has been particularly interested in being able to locate any given spot on earth, for such a capability would allow it to direct its missiles with more devastating accuracy. One leading astronomer asserted that geophysicists can presently locate surface features with an accuracy of ten centimeters and are developing measurements as accurate as one centimeter.

The various constraints on possible flight paths and the interaction between flight paths and mission objectives require space systems analysts to be well versed in celestial mechanics as well as engineering. Those responsible for space flights have to have precise descriptions of spacecraft motions in various situations in order to plan and manage missions. Scientists were asked to devise new tools of analysis to deal with the new data that were collected. For example, the use of two-way Doppler measurements of shifts in radio signals retransmitted by spacecrafts to earth allowed a more accurate calculation of orbits than had ever been possible before. New methods of analysis were also required to meet the needs of new applications. If scientists were to interpret photographs of the moon and planets from orbiting spacecraft and if they were to be able to measure surface features accurately, precise knowledge of orbits had to be assured.

The number of astronomers specializing in celestial mechanics and astrometry has remained small, probably about 6 to 7 percent of all astronomers. Nearly two-thirds are in universities, equally divided between teaching and research. Most of the remainder are employed by the federal government. Not many of the scientists interested in geodesy resided or worked in universities. Government agencies, including the Defense Department, have long provided the major support for and employment to most of those who worked in this area.

The new and added demands for advanced techniques and inventive application required additional trained people, for the number of specialists before the late 1950s had been extremely small. The universities did not greatly add to their staffs; and astronomers hardly considered the revitalization of geodesy as a major contribution to their science. Although they had once been closely identified with geodosy, most astronomers probably consider it to be a subdivision either of engineering or of geophysics.

Space explorations resurrected another group of studies, clearly astronomical, that had languished for years. In the last decade it has provided major fields of scientific research and experimentation in astronomy. By the 1920s astronomers increasingly turned their attention away from lunar and planetary study. Some techniques even then available were left unexploited and unused. With little additional information available then or in the foreseeable future, lunar and planetary studies withered. Solar studies continued, but the difficulty of observing the sun through an obscuring atmosphere greatly hindered research. One astronomer estimates that probably no more than 2 percent of astronomers in 1958 were interested in planetary studies.

One of the first questions to be answered by space experiments was the number of meteoroids in space and thus the degree of danger to satellites and men in spacecrafts. The number of objects capable of causing serious damage was far less than the worst predictions made by scientists. Later, longer missions allowed scientists their first close look at the moon. The detailed pictures of the surface by the Ranger and Lunar Orbiter satellites and Surveyor landers gave scientists an extraordinary flood of data about the earth's nearest neighbor. One astronomer deeply involved in planetary studies described to the author in an interview how he feels about the situation: "There has been a revolution in planetology due to NASA's contribution. We now have a separate division of the American Astronomical Society that is concerned with planetary science. Four or five years ago, it amounted to almost nothing — few papers to present. In our most recent meeting we had lots of papers. It is a real revolution with lots of research and all kinds of astronomers interested. . . . For a long time astronomers ignored the planets. With what information we had there weren't many difficult problems to solve. Study of the planets was a game that second-rate astronomers played in."

The judgment may be too harsh since some able astronomers had involved themselves in the study of the planets; but clearly, the new information gathered through space missions provided greater opportunity for scientific development than ever before.

Another leading astronomer pointed out that the extensive lunar experiments of the Apollo program, and those that have allowed more limited space probes of the surfaces of Mars, Venus, Jupiter, and Mercury, have provided enough information to make planetology a full-fledged, complex field in its own right. Moreover, scientists can engage themselves in studies of the planets. They provide a tool for identifying common features among the planets, and assisting in tests of causal relationships.

The observations to date have allowed astronomers to answer a number of first-order questions such as: Are the far and near sides of the moon the same or different? Do the planets have atmosphere? Scientists can proceed to second- and third-order questions that ask about the composition of the atmosphere and the geographic, geologic, and chemical differences among the planets. It is increasingly clear that the physical structure of the moon and of Mars and Mercury are somewhat different from that of the earth. When astronomers more fully understand the causes of the differences and have mapped the surface of Venus and probed its hot, dense atmosphere, they may hold the key to an understanding of the origin of the solar system.

More directly relevant to issues that affect our daily lives, astronomers may be able to contribute valuably to knowledge about the earth. They should soon be helping geologists explain how and why mineral deposits form, forecast earthquakes, and comprehend the processes of mountain building and continental movement. The enlivening of lunar and planetary studies will then be judged as much more than merely the refurbishing of some old branches of astronomy; it will also be a practical step toward improving our living conditions on earth.

To date NASA has accomplished much more than simply bringing fresh breaths of information and data into the moribund studies of the moon. It has also transformed the style of the science and the activity of the scientists. They no longer need merely *observe* the moon, for the astronauts brought back samples of rock and soil to examine and with which to experiment; they also set up sensor instrument arrays — for example, seismic recorders — from which information is still being received. Lunar astronomy and planetary astronomy for the first time are becoming *experimental* sciences.

Observation of the sun from space has proved to be as rewarding as astronomers interested in solar radiation had expected, opening the field of solar studies to new and exciting development. Although the links between solar radiation and meteorological phenomena are not clearly known, scientists are beginning to investigate possible interactions. They hope to be able to improve prediction of the effects of solar activity on radio communications. The Skylab studies produced a flood of data, indicating that the sun's corona is much more dynamic than earth-bound observations had led scientists to believe; furthermore, flares, solar prominences, and other short-lived solar events are more common and diverse than expected. What has seemed from earth to be minor happenings on the sun can be major events when seen from space in full spectrum. Years of former ground-based data on solar activity may be recalibrated to be made more useful in explaining the sun's behavior.

The enlivened branches of astronomy — lunar, planetary, and solar — are uniquely dependent upon space observation. The scientists who have specialized in them need satellites and orbiting laboratories to carry out their research. Only NASA possesses the capability that allows such research to proceed; it is not surprising therefore that NASA supports 70 to 80 percent of all planetary research.

The Revelation Effect: NASA's Creation
of New Astronomical Fields

Among the exciting and unexpected consequences of space exploration, which could hardly have been guessed beforehand, has been the discovery of new objects and phenomena in the solar system, the galaxy, and beyond. Dr. Jesse Greenstein has written that since the recent discoveries have been made in the hitherto hidden ranges of the electromagnetic spectrum, "it [is] clear that the astronomical universe was in many ways still largely unexplored" (National Academy of Sciences, vol. 2, 1972, p. xiv). And Drs. Franco Pacini and M. J. Rees (1973, pp. 98-105) point out that

> ... the universe revealed by astronomy over the past three centuries has been a calm one, except for the occasional flare-up of a nova. This view has been *shattered* by the discovery of phenomena that involve rapid and violent changes. (Emphasis added.)

Space experiments that revealed the universe of high-energy radiation in all its magnificence did not receive immediate and large support by NASA. Both scientists and the managers of the manned space program were primarily interested in the conditions of space around the earth and moon. It is not surprising that experiments designed to test theories about the space environment, through which satellites and spacecraft would soon fly, took priority over speculative probes looking for unknown, possibly unimportant, phenomena.

The first American satellite, Explorer I, in 1958 revealed a deep zone of radiation girdling the earth. It was named the Van Allen Belt, after the principal investigator whose experiment had recorded the phenomenon. Later probes and observations by satellites provided detailed descriptions of the earth's magnetosphere that produces the radiation belts and of their complex interactions with the solar wind.

The early discoveries led to the development, in the early 1960s, of a new astronomical specialty: particles and fields. This specialty is concerned with the properties of atoms, nuclei, and electrons (particles), and electrical and magnetic fields that are present in space. This new field of research is important to astronomy for it allows what is still rare in the science, "the possibility of measuring directly, *in situ*, many of the most important astrophysical processes difficult to study experimentally on the ground (National Academy of Sciences, vol. 2, 1972, p. 122). Before this field was opened for study some astronomers doubted the existence of the solar wind, the lower levels of the cosmic-ray nucleon spectrum was unknown, and the cosmic-ray electrons had not been discovered.

The new studies of particles and fields have contributed greatly to astronomers' understanding of stellar processes, but they promise as well to help scientists analyze the nature of laboratory plasmas and to provide valuable clues in attempts to produce controlled thermonuclear fusion. In the case of particles and fields, the technological needs of space exploration, manned or unmanned, and the interests of astronomers converged; both benefited, and from the knowledge they gained, the general public may reap rewards as well.

Space, or at least very high-altitude, observation has allowed scientists greatly extended study of infrared and ultraviolet radiation, both supplementing and improving data provided by ground-based instruments. Atmospheric water vapor absorbs infrared wavelength radiation in the 25 to 300 μ, the middle and shorter ranges; yet the most powerful discrete sources of electromagnetic radiation in the universe and in our own galaxy emit the bulk of their energy in these ranges. Some of the objects emit more energy in the infrared than in all other wavelength regions combined. These unexpected and still unexplained results deserve and are receiving close attention from astronomers; and NASA has provided strong support for the study of and research into infrared radiation.

The Orbiting Astronomical Observatory-2 (PAS-2) opened a new era in stellar-space astronomy, allowing for better viewing of ultraviolet radiation than that attained through the Aerobee-Hi sounding rockets. One of the most important astronomical discoveries was that of molecular hydrogen in interstellar space. The observations indicate that some stars lose substantial mass as gases steam away from them at escape velocities, appreciably modifying the surrounding medium.

It is the space observations of high-energy radiation, X rays and gamma rays, however, that promise the greatest change in astronomical theory and the deepest understanding of the universe. What that change may be and what its effects may be upon astronomy, physics, and other sciences and the lives of Americans cannot now be predicted. A prominent scientist, Dr. Freeman Dyson of the Institute of Advanced Studies, commented (Metz, March 16, 1973, p. 1114) that to ask what effects the new discoveries will have is like

> . . . asking Galileo, one year after he first turned his telescope on Jupiter, what would be the future of astronomy. This work is by far the most exciting thing to come out of the space program, scientifically speaking. We have found a whole menagerie of X-ray objects, perhaps 10 or 12 different classes. . . . For the firs⁺ time since Einstein, people are truly making progress in understanding what his equations say.

Another scientist also expressed himself with superlatives when describing the finding of the investigators whose X-ray experiments were first carried on satellites (Metz, March 2, 1973, p. 884): "They have had a greater influence on astronomy in the first few years than the 200-inch telescope on Mt. Palomar did."

Ricardo Giacconi and Herbert Gursky first proposed an experiment to look for X rays reflected from the moon. Whether they were truly interested in reflections that would have been very weak at best, or whether they offered that objective as a way of looking for X-ray objects elsewhere is not clear. Such data as were available to NASA officials and their scientific advisory panels did not recommend the experiment as a fruitful one. The advisers and officials may also have been a bit more cautious in this case since Giacconi was not a university researcher but, rather, was employed by a private laboratory.

When Giacconi's experiment was at length approved and flown in 1962, the detectors mounted on a sounding rocket did discover some remarkable X-ray sources in the regions of Scorpius and Cygnus. Other rocket-borne experiments continued over succeeding years. However, in the first half of the 1960s, the accumulated observing time of X-ray sources by all the rockets flown totaled only about one hour. Not until the Small Astronomical Satellite-A, usually known as Uhuru, was flown in 1970 did astronomers realize the promise of X-ray observations. The powerful X-ray emitter Cygnus X-1 was quickly identified, and by 1973, 125 separate sources had been found. A number of the sources were found to be orbiting other stars, and some were found to be of an unusual nature, such as the X-ray pulsars that are gaining energy. Other objects such as Cygnus X-1 may be the hard-to-believe stars, "black holes," that do not emit radiation.

After Uhuru's launching, another satellite, an Orbiting Astronomical Observatory (OAO), joined it. The OAO has been making successful observations of X-ray sources, too. The two satellites have provided astronomers with some of the most remarkable discoveries to date from the space program. Continued observation of this totally new and significant branch of astronomy is completely dependent upon a space program. Only an organization that can provide the capability to build, launch, and monitor satellites offers the means to maintain it.

A high-priority program in NASA and for astronomers is the launching of high-energy astronomical observatories (HEAOs). These satellites will be designed to detect X rays and other high-energy radiation; they are especially equipped to discover faint X-ray sources and to measure their characteristics. In its 1973 report to Philip Handler, president of the National Academy of Sciences, the Astronomy Survey Committee (National Academy of Sciences, 1972, vol. 1, p. 88) noted that

> . . . a measure of the importance attached to X-ray astronomy by astronomers is that they have scheduled large blocks of time on major optical instruments to exploit the discoveries and positional measurements of new X-ray sources by the Uhuru X-ray satellite.

The committee implicitly recognizes that the creation of a new, exciting field of astronomy is going to affect not only theories and ideas, but also the operations, organizational structures, and scientific relationships in astronomy. Old and new fields — old and new observing techniques as well as old and new skills — must work together as never before.

The Synergistic Effect: NASA's Contribution to Optical and Radio Astronomy

Space exploration is changing the style and organization of astronomy because the new modes of observation complement and supplement ground-based observation in complex ways that can enhance the effectiveness of both. There is little doubt that optical observation will remain central to the science. Optical instruments provide abundant basic information about stellar objects, offer greater resolution than either radio or high-energy detectors, and are more flexible, convenient, and available. Astronomers can also carry out ground-based observations at far lower cost than space observations. A major offsetting factor is that space observations offer unique opportunities to detect a wide range of phenomena. Astronomers must therefore work out the tradeoffs between the two observational modes — the one unique and costly, the other flexible and cheap.

The Astronomy Survey Committee (National Academy of Sciences, 1972, vol. 1, pp. 87-88) recognizes that

> . . . the spacecraft will play an essential role in the future of astronomy. X-ray astronomy will increasingly become a partner to optical and radio astronomy as more X-ray sources are identified and their properties are correlated with those of other wavelength bands.

Each will provide astronomers with information and data that will enlarge the meaning of the other. For the same reason, astronomers have recommended a large optical telescope in space — from sixty to 120 inches in diameter, depending on available funds. Such an instrument, above the atmosphere, could detect radiation through the entire ultraviolet and infrared range not accessible from the ground, as well as visible light. It would also offer a five-magnitude gain in sensitivity over ground-based instruments of comparable aperture and a good deal more observing time. Although it is expensive to operate, careful, selective use of such an instrument would return benefits similar in magnitude to those of HEAOs, revealing special events to be examined in other detail by ground-based instruments, or analyzing special phenomena beyond the limits of ground-based observation.

Two minor examples illustrate how space observations can enhance ground-based studies. During the voyage of Mariner 9 to Mars, ground-based astronomers kept a close watch on the planet. Before the spacecraft reached Mars, a trip of several months, those monitoring it from earth discovered a fierce dust storm that entirely enveloped Mars, obscuring its surface. Had the ground-based spacecraft experimenters watching the televised pictures of the planet not known that a dust storm was in progress, the information could not have been as easily interpreted.

R. M. Hjellming, a member of the staff of the National Radio Astronomy Observatory, reported another example. Space observers discovered a powerful X-ray source, Cygnus X-3, in 1966. Later ground-based radio observations showed that the source produced outbursts from time to time; an analysis of X-ray observations of the source by the satellite Uhuru revealed that the source emitted a double X-ray flux at a time when no radio data were taken. The X rays also normally varied regularly in a matter of hours, suggesting a double star. In early September 1972, certain radio astronomers in Canada and the United States discovered that Cygnus X-3 appeared to be one of the strongest point sources in the radio sky. Suspecting an event worth observing in all possible spectra, the astronomers notified radio astronomers round the world who tuned in at various different frequencies. At Mt. Palomar, astronomers began observing the source in early October with the 48-inch Schmidt camera and the two hundred-inch telescope at optical wavelengths, including infrared. Coincidentally with the radio observations, the X-ray instruments in satellites also observed the source.

Hjellming comments (1974, p. 1089) that there "may have been more investigation of a single object, by more astronomers, in a shorter period of time, than ever before in astronomy." The combined observations enhanced each other, for when astronomers plotted all of the information from different instruments and wavelengths, the data revealed much additional knowledge about the source. For infrared astronomers, the source turned out to be the first infrared object showing probable binary characteristics; for X-ray astronomers, the source is the shortest period X-ray binary known, by a factor of 10; and it is only the second source showing coupling between X-ray and radio events. From such new and puzzling data, astronomers have much to fuel theoretical analyses and guidelines for other and further observations.

Officials of NASA have long recognized that ground-based and space observations will advance astronomy optimally and provide the greatest returns of data and information by taking advantage of the synergistic benefits of the two observational modes. They have attempted to balance the needs of both in a jointly useful astronomy program. The high costs of space exploration have demanded the larger share of NASA's budget, but NASA has not ignored ground-based astronomy. From fiscal 1966 to fiscal 1972, NASA contributed an average of more than $9.3 million a year to ground-based astronomy, 17.6 percent of the total federal support provided for it (National Academy of Sciences, vol. 1, 1972, p. 65).

Since 1958 NASA has supported the building of new, or the refurbishing of old, instruments so that astronomers at ground observatories could meet the added demands made, and exploit the new opportunities afforded, by the space program. It has contributed support for many optical telescopes, ranging in size from 24 to 107 inches, as well as a number of large tracking instruments. It has also supported the construction of a number of solar-optical telescopes as well as solar-radio telescopes. In addition, it has contributed in full or in part to the building or updating of about a dozen radio telescopes (National Academy of Sciences, vol. 2, 1972, pp. 388-93, with authors' caution).

The peculiar observing conditions above the atmosphere required NASA to develop new instruments or adapt existing techniques for astonomical observing in space. Since information from satellites usually could not be physically delivered to earth-bound observers, it had to be transmitted by radio and television. For these purposes NASA devoted considerable effort and resources to the development of sensitive, reliable TV sensors. Photomultipliers fit the needs and limitations of satellite observation especially well. By the mid- to late 1960s, ground-based observatories had borrowed the new detectors, both the improved TV sensors and the photomultipliers. They have increased the efficiencies of optical telescopes significantly, for they can measure light accurately within 0.1 percent accuracy, while photographic plates are accurate within the 1 to 10 percent range only.

Several astronomers at large observatories believe that NASA's general support of astronomy both enabled and encouraged astronomers to experiment more boldly than ever before with new instruments that have enhanced the detecting range of existing telescopes. The easy availability of funds compared with earlier times had loosened budget constraints, allowing contemplation of new devices. And NASA's support for instrument development and adaptation of military equipment, such as improved infrared detectors and high resolution cameras, provided exciting new kinds of detectors. Because NASA pushed the technology of computer monitoring in real time, it may also have contributed to more effective observations with the large, ground-based telescopes.

There exist only incomplete and sketchy data on NASA's contribution to improvement in astronomical technology. The judgment of a number of NASA managers and scientists acquainted with the astronomical program and of several astronomers interested in instrumentation is that NASA has been a significant contributor. In 1969-70 the federal government accounted for 28 percent of the manpower devoted to general-purpose instrument development (National Academy of Sciences, vol. 2, 1972, p. 314). Most of the rest of the manpower used for that purpose was in the universities. It is reasonable to assume that NASA's share of the funds supporting that manpower was larger than any other agency's.

Astronomy as a "Big Science"

The space program has helped add to the fundamental knowledge of astronomy; it has opened new fields of inquiry in the science and revitalized old ones; and it has enhanced the older methods of observations. As meaningful as these contributions have been, the effects of NASA and its space program on the organization and structure of astronomy may equal or surpass them. The large flow of funds into the science has helped it move toward — if it has not already created — a Big Science status. It was estimated in 1972 that the costs of proposed new instruments, such as a large space telescope, would grow from a level of about $20 million a year at the beginning of a decade of development to around $200 million a year at the end of the decade. Total costs over the entire period of fabrication and launch would be at least $1 billion (National Academy of Sciences, vol. 1, p. 106). These figures are large enough to justify classing space astronomy as a Big Science.

The federal government already spends on astronomy an amount comparable to that spent by it on high-energy physics, the nation's most easily recognized Big Science. In 1969 the federal government obligated $282 million for basic research in astronomy, nearly 12 percent of the funds made available by the government for all basic scientific research. By 1972 the obligations for basic research in astronomy had declined by a third to $189 million, only 7.4 percent of the federal total

(National Academy of Sciences, vol. 2, 1972, p. 352; and National Science Foundation, 1973, p. 35, Table B-8). Of the smaller amount of funds allocated for astronomy in recent years, four-fifths or more has come from NASA. This and the larger amounts that were provided earlier have been sufficient to change the organization of astronomy from its patterns in the 1940s and '50s.

Dr. Hugh Dryden, deputy administrator of NASA from 1958 to 1965, early pointed out how the requirements of space science would affect the way astronomers work and manage their field (*Proceedings*, vol. 1, 1962, p. 89):

> The university scientist who participates in satellite and space-probe experiments finds an environment different from that to which he has become accustomed. Traditionally, a scientist conceives an experiment, builds the apparatus himself or has it built under his supervision in the university shop or by contract, carries out his own experiment, analyzes the data, and publishes his results. This relatively simple procedure is not possible in satellite and space-probe experiments, although a fair approximation to it is feasible for experiments with small sounding rockets. Satellite launching requires large rockets, special sites, a worldwide tracking and data-acquisition network, sharing by many experimenters in a single flight, and a large team of cooperating specialists. The scientist becomes involved in scheduling his work to meet a flight date once that date has been set. His apparatus must be engineered to meet several environmental requirements of vibration, temperature, exposure to radiation and charged particles, and so forth. Some universities are able to provide this service; others must depend on industrial help. Thus the role of the university scientist often *reduces* to concept of the experiment, development of laboratory prototypes of the equipment, analysis of the data and publication, *plus participation in a large team* to design the actual satellite, launch it, and receive the data. (Emphasis added.)

Despite the high expenditures on space experiments, the number of astronomers who have conducted space satellite experiments is still small. Less than 15 percent of astronomers participated in space projects in 1969-70, although space experiments have wider applications within astronomy than the number of astronomers directly involved indicates (National Academy of Sciences, vol. 1, 1972, pp. 60, 62). But Dr. Dryden's description of the complex team effort required to accomplish satellite missions will become a reality for increasing numbers of astronomers if and when more of them become involved in space observations.

Even before space requirements have been widely felt among the astronomical community, it had already turned to more complicated managerial organization of facilities than were known before World War II. To provide optical and radio instruments to a wide section of the increasing number of astronomers, a consortium of universities, the Associated Universities for Research in Astronomy, established the Kitt Peak National Observatory in Arizona, funded by the National Science Foundation. The observatory is open to all qualified users, and in 1971 astronomers from 33 small institutions used its facilities as well as astronomers from large universities or other observatory groups.

To furnish better radio astronomy facilities, a number of universities also banded together as the Associated Universities, Inc., to found the National Radio Astronomy Observatory. It has two centers, is funded substantially by the National Science Foundation, and is also open to all qualified scientists. Thus astronomy has been moving in the direction in which space science will take it still further. Increasingly, astronomers will be but a part of a team, established to serve their engineering, technical, and managerial needs along with their scientific interests.

The Astronomy Survey Committee warned that "for space projects . . . [NASA] centers are essential to the engineering of much of the flight hardware. Few complete satellites systems can be handled by university groups" (National Academy of Sciences, vol. 1, 1972, p. 114). And if the nation should proceed with development, launching, and use of the large space telescope, the astronomical community will have to mobilize itself to undertake so vast a program. The committee (vol. 1, p. 106) realizes that

> . . . a program of this magnitude requires the highest quality scientific leadership and the most advanced space engineering available. The highest quality scientific leadership in this field can be found in the academic community, and the highest degree of space engineering talent exists in the [NASA] centers. . . . Therefore, the best chance for success lies in the merging of academic talent with that in

the NASA centers. We suggest that NASA select one or more centers to carry out the engineering phases of the program and the National Academy of Sciences encourage the formation of a new corporate entity representing universities with strong programs in space astronomy.

Over the span of but one generation, astronomy has been transformed, first by the technology of radio observation, and now by space exploration. Once it was the preserve of individual scientists pursuing independent studies in universities and remote observatories. Astronomers lived and worked in considerable isolation from other sciences and with instruments and equipment left largely unchanged for more than sixty years. Today astronomers increasingly must work as members of teams on complex experiments that may involve the coordination of radio and optical observations with simultaneous space observations. Some of the scientists must assume the heavy administrative duties of managing the activities of the large astronomical observatories and centers, and of coordinating their interests and decisions with those of the universities and government agencies that provide support for astronomy. Astronomers participating in space experiments often find they have to concern themselves with complex engineering tasks, meet large financial responsibilities, and manage sizable personnel staffs. Such a radical shift in the demands made upon astronomers will doubtless greatly change the kind of manpower found in the science.

The rise in funding and the opening of opportunities for new experimentation have indeed increased the quantity of manpower. Through the 1960s and into the 1970s, there was an unprecedented increase in the number of astronomers. By 1970 the total number had approximately tripled in a decade, to about 2,500, averaging an annual growth rate of nearly 15 percent over the latter part of the 1960s. The number of scientific and technical personnel in astronomy increased from about twelve hundred in 1959-60 to around 3,200 in 1969-70 (National Academy of Sciences, vol. 1, 1972, p. 56). The number of doctorates in the science increased at the same rate (vol. 2, p. 325). These growth rates were about four times larger than those over the previous forty years.

Through its project grants and sustaining university predoctoral traineeships and other university programs (1966 and '67), NASA supported almost 40 percent of the 66 new Ph.D.s in astronomy (National Aeronautics and Space Administration,

1968, p. 29). Even before NASA support became an important means of attracting students to the field, the number of doctorates in astronomy had begun to rise sharply. The development of radio astronomy and support of astronomy by the National Science Foundation and the Armed Services contributed significantly to that rise. No doubt the general stimulation of space exploration, aside from NASA's educational aid, attracted some students.

With the cutbacks in the space program and the leveling off of government funding for all sciences, including astronomy, it now appears that NASA may have induced a larger inflow of astronomers than is now needed at current and foreseeable government funding for the sciences. By 1966 NASA engaged the services of from eighty thousand to a hundred thousand engineers and scientists, 5.5 percent of the national total (Newell, 1967, p. 1). By 1968, when NASA reached its peak, its managers realized that the figures would be even higher. Some overly optimistic forecasts suggested that by 1970 as many as one-fourth of all scientists and engineers would be working on space programs. Although NASA officials doubted the accuracy of such a forecast, they certainly did not want to be accused of draining the nation of its highly trained, skilled scientific and technical work force and creating shortages for other sectors of the economy (see statement by Dr. Hugh Dryden in National Aeronautics and Space Administration, 1962, p. 90).

Dr. Thomas L. K. Smull, director of the Office of Grants and Research Contracts, explained the view of NASA's managers in 1962. He declared that NASA needed the wholehearted cooperation of the universities for they supply "the creative ideas upon which much of the program is based," and they also develop "the scientific and technical manpower that will be required for successful conduct of the program." He assumed that "the university is the only segment of the team undertaking the space program that produces manpower. The other two partners in this enterprise — industry and government — only *consume* manpower" (National Aeronautics and Space Administration, 1962, pp. 51, 53). His assumption is probably erroneous. An employer does not necessarily consume manpower; if the labor force has any chance whatsoever to improve its skills and increase its knowledge about and understanding of the tools and techniques in use, it will increase its quality, adding to the productiveness of manpower. On-the-job training is probably as important a factor for improving productivity as schooling, even in the advanced sciences.

The National Aeronautics and Space Administration did not need to support Ph.D. students in the numbers it did, unless its officials believed that the nation could long sustain the growth in number of space scientists that was experienced in the early 1960s. The numbers of astronomers increased per year at 10 to 15 percent, high enough to double the ranks in five to seven years! One might argue that present government support for science students, scientists, and basic research is too low for the national good. It seems hardly possible that taxpayers would support the research and development at the rate they did from 1953 to 1967 — 10.3 percent. Continued for another fourteen years, research and development spending would have been nearly a third larger than total expenditures for national defense in 1967.

The supply of scientists may well be more flexible and their numbers more expansible in any particular area or field than lay citizens, or even managers, recognize. Scientists can and do slide from specialty to specialty without great difficulty. Any labor force of well-schooled persons provides a reservoir of flexible manpower capable of easily and quickly transferring skills and competencies to related occupations. In the case of astronomy, both the new direction of research (toward particles and fields, and X-ray experiments) and the type of research (team efforts involving complex engineering requirements) attracted physicists. So many were attracted that by 1972 the number of "astronomers" with doctorates in physics exceeded for the first time the number who earned doctorates in astronomy (National Academy of Sciences, vol. 2, 1972, p. 323). A smaller portion of other Ph.D. researchers, around 6 percent, left their specialties, such as chemistry, engineering, and geology, to become astronomers too (vol. 2, p. 334, Table 9.18).

The rapid inflow of manpower to the field probably lowered the average age of astronomers. In 1968 nearly 35 percent of all scientific and technical personnel were under the age of thirty and 37 percent were in their thirties; thus almost three-quarters of these men and women were younger than forty (National Academy of Sciences, vol. 1, 1972, p. 59). No data are available for earlier periods.

The increase in numbers of astronomers has been so large and so rapid that today's astronomical community hardly resembles that of a decade ago, and it is indeed different from that of a generation ago. It is a far more open science, with more numerous facilities and research opportunities; it is populated with young scientists and technicians who come from more diverse educational backgrounds than ever before. In this change, NASA has been a principal cause. Its large expenditures — more than $100 million annually for basic research alone — dwarf the funds astronomy once had for its modest work in the 1940s and early '50s. The funding provided by NASA also dominates the federal support now provided for astronomical research. In 1970 NASA contributed 65 percent of federal research grants for astronomy in academic institutions and 56 percent for all astronomical institutions (National Academy of Sciences, vol. 2, p. 351).

Secondary Consequences of Astronomy's Transformation

With its transformation, astronomy has become a complex scientific field — or more accurately, a group of related but rather distinct sciences. Each has as many or more personnel involved in its conduct as the whole of astronomy a generation ago. Each now possesses a complicated array of instruments and equipment that can best be used to advantage when operated cooperatively with those of other astronomical areas. Financing depends primarily upon the federal government, with the largest share coming through NASA, whose allotments for astronomy make up less than 5 percent of its total budget. The federal government distributes its funds through NASA research centers, national observatories, universities, and other nongovernment corporations, each of whom must maintain administrative staffs to manage operations and financial flows. The varied astronomical areas of research, the large number of astronomers, and the many adiministrative agencies require more effort and impose far higher transaction costs than in earlier times; it is more difficult and more complex to set priorities, to ensure joint or coordinated efforts, and to manage effective programs.

One administrator of astronomy programs has pointed out (in an interview with the author) that the problems of setting priorities in the science will grow. "Astronomy," he said, "is beginning to fractionate. We find solar physics, planetary studies, and research into the interplanetary medium, each, as a separate discipline in addition to stellar and galactic astronomy. They were all within the purview of a single science only a few years ago. But the information flow has become so great and specialized that we can now concentrate in distinct specialties of worth."

The fractionation has gone so far that in reporting on scientific advances, the American Association for the Advancement of Science separates earth and planetary studies from physics and astronomy (*Science*, December 1973). Such labeling also suggests more than a simple reorganization of astronomy into subfields, but also a combining of various parts with other sciences. Astrophysics received equal billing with astronomy in the title of the recent report by the prestigious Astronomy Survey Committee. Dr. Harvey Brooks for the Committee on Science and Public Policy of the National Academy of Sciences notes the "increasingly close relationship [of astronomy] to virtually every subfield of physics and, increasingly, to chemistry" (National Academy of Sciences, vol. 1, 1972, p. v).

Despite the separation of fields in astronomy and new combinations, the Astronomy Survey Committee succeeded, in its 1972 report, in setting straightforward priorities for future astronomical programs. One commentator observed that "few government advisory committees have been able to accomplish" such a task. He suggested, however, that the committee made agreement on priorities easier than it might have by opting for expensive space programs without enough attention to these programs' past or likely future scientific effectiveness (Metz, 1972, p. 248).

If NASA's budget is large enough, astronomers may reasonably hope that funds will trickle through and around the specific space programs to support a condsiderable amount of other kinds of astronomical research. Thus the present unanimity of astronomers may be more a function of NASA's single large contribution to the science than to a common scientific interest and outlook among the diverse specialties that astronomy now includes. Should the government distribute its funds through a number of agencies, the problem of securing the scientists' agreement on priority may become far more difficult than at present.

With recent cutbacks in NASA's budgets, astronomers see some difficult problems facing them. First is the balance of resources between the national observatories and the university-based astronomy departments. Astronomers in the latter may be at a disadvantage in developing new instrumentation. In optical observation, an area they have dominated in the past, they have made major technical advances; they fear their contributions may be diminished in the future. Should an increasing share of the national funds allocated to astronomy go to national observatories and government in-house programs, the universities may find that they cannot hold their place and continue their influence in the science. A kind of nationalization of facilities and observation will follow the nationalization of financing.

In its heyday of expanding budgets, NASA encouraged a number of colleges and universities (that had been involved) to build up space science departments. As budgets shrink, NASA officials — and the astronomers who advise them — must decide not only the relative preference in support between university and national observatories but also the priority among schools. When confronted with the latter issue in their science, the Physics Survey Committee (with Dr. Alan Bromley as chairman) explicitly recommended that scarce resources should be concentrated on "the support of excellence" at major facilities and at the best universities, rather than be broadly distributed (National Academy of Sciences, August 1972). The committee members were prepared to recognize that under such conditions the science of physics will henceforth be an elitist science.

Astronomers at Harvard, Massachusetts Institute of Technology, and Princeton, reported that astronomy is implicitly following the same policy. Schools whose astronomy programs depended completely upon government funds have had to be reduced or terminated. The larger, stronger schools that have private resources to supplement the reduced federal funding have been able to hold their talented researchers and continue their astronomy programs, although sometimes on a smaller scale.

Even in the strong "elite" universities and private observatories, the space program can have a destructive, or at least morale-shattering, effect upon the scientists involved when the government cuts budgets or, to save funds, stretches programs over longer periods than originally planned. For example, in January 1973 NASA cut the HEAO program by $200 million, although the astronomical community had given it the highest priority for space exploration.

Professor Donald C. Morton, Princeton University Observatory, saw NASA's action as a serious breach of trust (quoted in *Sky & Telescope*, March 1973, p. 143):

> Many young people have based their studies and careers on the HEAO and other government-supported projects and are now going to have to re-think their futures. Who among them, and us, will plan on devoting the next ten years of our lives to such a project if the government does not honor its promises?

The wide swings in government funding and consequent sudden shifts in program plans add to what already are sizable risks for scientists who conduct space experiments. The Princeton astronomers became involved with a major satellite project in 1959; the satellite was planned to fly in 1965, but because of technological difficulties, was not launched until 1972, seven years later. At that, the experimenters were more fortunate than others who worked on satellites that failed in launching or soon afterward.

The risks of long lead times are compounded by the large commitment a space project requires. One knowledgeable astronomer, during an interview with the author, said, "You are going to get little else done over the years you work on it, and even then you do not know if there is going to be much of a payoff. You could wind up with very little research." Another scientist, an astrophysicist, told the Subcommittee on Space Science and Applications (U.S. Congress, House, 1973, pp. 15-16):

> . . . [T]he prime scientific resource in our country [the scientific researchers] should be treated tenderly. . . . [T]o be sure, the scientific community will bounce back for the most part, but I think that the most talented experimenters, who discover the really new things, may well become disillusioned and will work in other fields.

Perhaps enough university researchers will always be found to contribute sufficient time and talent to space science to help the nation maintain a program of quality; it is not clear whether the demands and risks of space science will allow present relationships between the universities and the government centers, national observatories, and space agency to continue.

The transformation of astronomy effected first by radio astronomy and later by NASA's massive expenditures relative to past levels, and its opening of space, is but partly accomplished. The old structure of the science and the ways it was organized and managed have been swept away; present forms and relationships may be temporary and in flux. The National Aeronautics and Space Administration and its managers have helped create a new science, more complex and costly than the former science, with a new and larger, broader constituency of scientists. Astronomers, even with all of their supporters and amateur star-gazers, are not apt ever to be a potent political pressure group. But in creating today's astronomical community, NASA has assumed a more difficult responsibility than officials may have realized in the early days of the space program.

References

Hjellming, R. M. "An Astronomical Puzzle Called Cygnus X-3." *Science* (December 14, 1973), vol 182.

Metz, William D. "Astronomy from an X-Ray Satellite." *Science* (March 2, 1973), vol. 179.

_____. "Physics at a Turning Point?" *Science* (March 16, 1973), vol. 179.

_____. "Report on Astronomy: A New Golden Age." *Science* (July 21, 1972), vol. 177.

National Academy of Sciences. *Astronomy and Astrophysics for the 1970's.* Volumes 1 and 2. Washington, D.C.: National Academy of Sciences. 1972.

_____. *Physics in Perspective.* Washington, D.C.: National Academy of Sciences. August 15, 1972.

_____, and National Research Council. *Ground-Based Astronomy: A Ten-Year Program.* Washington, D.C.: National Academy of Sciences. 1964.

National Aeronautics and Space Administration. *A Study of NASA-University Programs.* Washington, D.C.: National Aeronautices and Space Administration. 1968.

_____. *NASA and the Universities.* Washington, D.C.: National Aeronautics and Space Administration. November 1, 1962.

National Science Foundation. *National Patterns of R&D Resources, 1953-1973.* Washington, D.C.: National Science Foundation. 1973.

Newell, Homer E. "Space Astronomy Program of NASA." In *Astronomy in Space*. Washington, D.C.: National Aeronautics and Space Administration. 1967.

Pacini, Franco; and Rees, M. J. "Rotation in High-Energy Astrophysics." *Scientific American* (February 1973).

Proceedings of the NASA-University Conference on the Science and Technology of Space Exploration. Volume 1. Washington, D.C.: November 1962.

Science (December 28, 1973), vol. 182.

Sky & Telescope (March 1973).

Swenson, L. S.; Grimwood, J. M.; and Alexander, C. A. *This New Ocean*. Washington, D.C.: National Aeronautics and Space Administration. 1966.

U.S. Congress, Senate. *NASA Authorization for Fiscal Year 1969.* Hearings before the Committee on Aeronautics and Space Sciences. 90th Cong., 2d Sess. 1968.

_____, House. Testimony of Professor Robert Hofstadter before the Subcommittee on Space Science and Applications. March 14, 1973. Mimeographed.

Chapter Four

●

Predicting the Unpredictable:
Economic Impacts of Weather Satellites

Meteorological satellites represent one of the most valuable technological advances in the history of meteorology. The launching of the television and infrared observation satellite, Tiros I, on April 1, 1960, revolutionized methods of weather observation. Tiros I demonstrated that meteorological satellites are an effective way to overcome the limitations of conventional observation methods. Previous techniques, such as radar, weather reconnaissance aircraft, weather ships, and weather balloons, supplied information for less than a fifth of the earth's surface; Tiros I covered almost the entire globe.

To assess the impact of Tiros I and its successors, one must examine two major relationships:

(1) The relationship between weather satellites and the accuracy of weather forecasts. Have weather satellites contributed to any observed improvement in forecast accuracy?

(2) The practical value of improved weather forecasts. What is the dollar value of the savings and benefits from better weather forecasting? What proportion of these savings is due to meteorological satellites?

In order to answer these questions, one must begin with an overview of the U.S. meterological satellite program. The following section is largely based upon the extensive historical descriptions found in Widger (1966, pp. 38-49) and Legislative Reference Service (1962, pp. 26-36).

The National Meteorological Satellite Program

National support for satellites did not emerge until after the Russians had launched their first Sputnik in 1957. In the aftermath of Sputnik, the Defense Department established the Advance Research Projects Agency (ARPA). The agency initiated many of this country's space projects before NASA was established. During the spring of 1958, ARPA determined that a meteorological satellite would be an appropriate part of the U.S. space program and requested the three military services (Army, Navy, and Air Force) and the U.S. Weather Bureau to furnish their recommendations in support of such a project. At that time the national clearinghouse for weather forecasts was the Weather Bureau, which has undergone two major reorganizations since 1965. From July 1965 through October 1970, it was part of the Environmental Science Services Administration (ESSA). In October 1970, that agency was reorganized into the National Oceanic and Atmospheric Administration (NOAA). Beginning with the Environmental Science Services Administration, satellite activities were set up in a separate entity now known as the National Environmental Satellite Service (NESS) of NOAA. Both reorganizations were designed to consolidate federal activities in the environmental area into a single agency (Hughes, 1970; Popkin, 1967; and Space Science and Engineering Center, vol. 1, pp. 399-434).

Representatives of the Army, Navy, Air Force, and Weather Bureau subsequently served on ARPA's Committee on Meteorology, which was formed to guide the development of the satellite that eventually became Tiros I. Project Vanguard, under the direction of the Naval Research Laboratories was already under way. Vanguard II was successfully placed into orbit in February 1959, but because of a variety of technical problems, no meteorologically usable data were obtained. However, experiences gained during the development and operation of Vanguard proved valuable in the design of Tiros.

Program responsibilities were centered in ARPA until April 1959, at which time these functions were transferred to NASA. Preliminary plans for the organization and funding of the national meteorological satellite program were first drawn up in 1960-61 by an interagency working group, the Panel on Operational Meteorological Satellites. The panel report in April 1961 recommended assigning management responsibility for a national operational meteorological system to the Weather Bureau, with NASA participating in the developmental aspects of the program. Under a 1962 agreement between the Weather Bureau and NASA, the former was to determine overall meteorological requirements, process satellite data for operational purposes, disseminate data and forecasts, and conduct research on the climatological use of satellite data. The latter (NASA) was to be the research and development organization with responsibilities for designing, building, launching, and testing satellites. When a satellite became operational, it was to be turned over to the Weather Bureau for continued research (Delbeq and Filley, pp. 72-85).

Eight Tiros satellites were launched between April 1960 and December 1963. The pictures provided by these satellites began to satisfy the meteorologists' need for worldwide picture coverage of the weather. Because local weather is often influenced within a relatively short time interval by remote atmospheric conditions, meteorologists had always sought closely spaced observation points with a maximum time interval between measurements of twelve hours. Traditional weather observation techniques could not satisfy this need because, even with radar and weather reconnaissance planes and ships, only a fifth of the earth's surface could be observed. Observations in the ocean areas were particularly inadequate.

In addition to supplying more complete weather coverage, Tiros satellites provided types of data that could not be obtained through other means of observation. The detailed pictures of cloud patterns provided by satellites cannot be duplicated by balloons, rockets, or radar. Satellites are also particularly suited for the observation of reflected solar and emitted infrared radiations. The continual exchange of energy between the earth and the sun produces winds, clouds, rain, and other forms of weather. Because observations of outgoing infrared radiation from the earth were impractical on any regular basis before satellites, most meteorological research and almost all weather forecasting techniques have been forced to assign values for the energy input and losses of the atmosphere.

While the Tiros satellites provided data never before available in any form, there were these problems: (1) only about 20 percent of the earth was visible to these satellites on any one day, and (2) not all latitudes could be photographed. Nimbus I, launched in August 1964, overcame these limitations by taking pictures of every part of the earth (except for the winter pole) once every 24 hours. Nimbus I also provided a breakthrough in the use of the automatic picture transmission camera. This slow-scan camera made it possible for relatively inexpensive equipment at ground stations to receive pictures showing the cloud cover over the surrounding area as it existed just 3.5 minutes earlier.

The second post-Tiros milestone occurred when the Applications Technology Satellite (ATS-1), the first geostationary satellite equipped with a camera, was launched in December 1966. Because of the particular orbit plane of the synchronous (or geostationary) satellite, it appears to be stationary over a particular point on the equator. Thus the spin-scan cloud camera on the synchronous satellite can continuously view the same portion of the earth. Cloud displacements over short periods of time can readily be measured, and from them many features of the wind field can be inferred.

The next major breakthrough came with Nimbus III, launched in April 1969, when vertical profiles of temperature, ozone content, and water vapor were furnished by satellite for the first time. Subsequent improvements in the Nimbus series resulted in more sophisticated spectrometric measurements of the temperature and moisture fields. In addition, satellites are now able to determine wind information at cloud level through the careful measurement of cloud motions.

The first operational weather satellite, ESSA 1, was orbited in February 1966. This was followed by eight additional first-generation (ESSA) operational satellites and a series of second-generation (NOAA) operational satellites beginning in December 1970, as shown in Table 4-1 (Hubert and Lehr, 1967, pp. 29-52; Popkin, 1967, pp. 154-60; U.S. Department of Commerce, 1971, pp. 1-5; and National Aeronautics and Space Administration, 1975, pp. 61-64, 134). These systems provide global observation of atmospheric elements to central readout stations; local weather information is transmitted via satellite automatic picture systems to local receivers distributed all round the world.

The major weather satellite development of the 1970s has been the launching of the synchronous meteorological satellite (SMS), which is a NASA-funded prototype for the geostationary operational environmental satellite (GOES). The SMS/GOES carries an infrared spin-scan radiometer system which provides nearly continuous observation of cloud cover day and night and measurements of cloud-top and surface temperatures. The first full-time weather satellite in synchronous orbit was launched on May 17, 1974.

Federal expenditures in support of the operational weather satellite program totaled $55.7 million in 1974. This funding represented 13 percent of total federal expenditures on meteorological operations. An additional $36.5 million was devoted to research supporting the weather satellite program. These research expenditures amounted to 53 percent of total federal expenditures on research supporting meteorological operations (U.S. Department of Commerce, 1974, pp. 5-55).

Advances in Weather Forecasting Techniques

Satellite observation is not the only technological advance that has helped transform weather forecasting during the last two decades. The development and refinement of mathematical models for weather prediction have been a major· influence on weather forecasting since the mid-1950s. The traditional method used by weather forecasters has been the synoptic approach, which is both empirical and largely subjective in nature. Using this approach, individual forecasters gather data concerning present and recent atmospheric conditions. They then use experience and judgment to extrapolate this weather pattern.

As weather forecasters became more aware of the limitations of synoptic forecasting, they intensified their search for methods which were less subjective. Although numerical weather prediction had been envisioned in the 1920s, its application was not feasible until modern electronic, high-speed, stored-program computers became available in the 1950s. In numerical weather prediction, a mathematical representation of an observed atmospheric state is developed. A set of time-dependent mathematical equations, which represent the physical laws governing atmo-

Table 4-1

U.S. Meteorological Satellite Launchings

(1960-72)

Date of Launch	Name	Launch Vehicle	Remarks
April 1, 1960	Tiros I	Thor-Able	First weather satellite providing cloud cover photography.
November 23, 1960	Tiros II	Thor-Delta	
July 12, 1961	Tiros III	Thor-Delta	
February 8, 1962	Tiros IV	Thor-Delta	
June 19, 1962	Tiros V	Thor-Delta	
September 18, 1962	Tiros VI	Thor-Delta	
June 19, 1963	Tiros VII	Thor-Delta	
December 21, 1963	Tiros VIII	Thor-Delta	First weather satellite designed to transmit continuously local cloud conditions to ground stations equipped with APT receivers.
August 28, 1964	Nimbus I	Thor-Agena B	Carried advanced videcon camera system, APT, and a high-resolution infrared radiometer for night pictures.
January 22, 1965	Tiros IX	Thor-Delta	First weather satellite in a sun-synchronous orbit.
July 2, 1965	Tiros X	Thor-Delta	
February 3, 1966	ESSA 1	Thor-Delta	First operational weather satellite; carried two wide-angle TV camera systems.
February 28, 1966	ESSA 2	Thor-Delta	Complemented ESSA 1 with two wide-angle APT cameras.
May 15, 1966	Nimbus II	Thor-Agena B	
October 2, 1966	ESSA 3	Thor-Delta	Odd-numbered ESSA spacecrafts carry two advanced camera systems. Even-numbered spacecrafts carry two automatic picture transmission camera systems.
December 6, 1966	ATS-1	Atlas-Agena D	Provided continuous black-and-white cloud cover pictures from a synchronous orbit, using a Suomi camera system.
January 26, 1967	ESSA 4	Thor-Delta	
April 20, 1967	ESSA 5	Thor-Delta	
November 5, 1967	ATS-3	Atlas-Agena	Provided continuous color cloud cover pictures from a synchronous orbit, using three Suomi camera systems.

Table 4-1 (continued)

Date	Satellite	Launch Vehicle	Description
November 10, 1967	ESSA 6	Thor-Delta	
August 16, 1968	ESSA 7	Thor-Delta	
December 15, 1968	ESSA 8	Thor-Delta	
February 26, 1969	ESSA 9	Thor-Delta	
April 14, 1969	Nimbus III	Thor-Agena	Provided first vertical temperature profile on a global basis of the atmosphere from the spacecraft to the earth's surface.
January 23, 1970	ITOS I (Tiros M)	Thor-Delta	Second-generation operational meteorological satellite.
April 8, 1970	Nimbus IV	Thor-Agena	Fifth in a series of seven advanced research and development weather satellites.
December 11, 1970	NOAA 1 (ITOS A)	Thor-Delta	Second-generation operational meteorological satellite.
August 16, 1971	Ecole (CAS-1)	Scout	French satellite to gather data from constant-density surface balloons relaying meteorological data for the study of the characteristics and movements of air masses. New balloons are released daily from three sites in Argentina for this cooperative French-U.S. project.
October 15, 1972	NOAA 2 (ITOS D)	Thor-Delta	Second-generation operational meteorological satellite.
December 11, 1972	Nimbus V	Thor-Delta	Provided the first atmospheric vertical temperature profile measurements through clouds.
November 6, 1973	NOAA 3 (ITOS F)	Thor-Delta	Second-generation operational meteorological satellite.
May 17, 1974	SMSI	Thor-Delta	First full-time weather satellite in synchronous orbit.
November 15, 1974	NOAA 4 (ITOS G)	Thor-Delta	Second-generation operational meteorological satellite.

Source: National Environmental Satellite Service, National Oceanic and Atmospheric Administration, U.S. Department of Commerce, *Satellite Activities of NOAA 1972* (Washington, D.C.: U.S. Government Printing Office, June 1973), p. 10; and National Aeronautics and Space Administration, *Aeronautics and Space Report of the President, 1974 Activities* (Washington, D.C.: U.S. Government Printing Office. 1975), p. 134.

spheric motion, is then used for purposes of prediction. The solution of these complex mathematical equations requires the computing power of the largest and fastest computers as well as the extremely advanced numerical methods (Maunder, 1970, pp. 266-89).

The application of numerical weather prediction techniques by the National Weather Service began at the National Meteorological Center in 1958. The center supports the National Weather Service by providing regional weather forecast stations round the United States with forecast guidance; and it acts in a sense as a forecaster's forecaster by providing the regional forecaster with information that he or she requires in the preparation of forecasts.

The accuracy of the initial mathematical models was limited by gaps in scientific knowledge and the existing state-of-the-art in numerical methods. Advances in both of these areas were prerequisites for progress in weather prediction. Throughout the 1960s and early '70s, computer capacity proved to be an important factor limiting further progress in numerical prediction. In fact, most major improvements in the accuracy of the National Meteorological Center's forecasts coincide with its acquisition of a new and more powerful computer system. Each new computer — the IBM 704 in 1958, the IBM 7090 in 1962, the CDC 6600 in 1966, and the IBM 360-195 in 1974 — enabled the center to develop more sophisticated and accurate mathematical models. During its seventeen-year experience with operational numerical weather prediction, the center has achieved impressive progress in improving forecast accuracy and extending the length of the forecast period (Shuman, March 1972).

Technological Advance and Forecast Accuracy

How have these technological developments in weather forecasting — satellite observation and numerical weather prediction — affected the accuracy of weather forecasts presented to the users of weather information? Two sets of data are available from the National Meteorological Center to answer this question. Comparative data document the superior accuracy of the center's man-machine mix forecasting system over traditional methods of forecasting. Trend data trace improvements in accuracy at the center over time.

A comparison of numerical prediction and subjective forecasts for varying forecast periods is shown in Figure 4.1. Correlation coefficients were computed between forecast and observed sea-level pressures for as much as six days. Four different forecast methods were used — persistence, unaided subjective, raw model, and model-aided subjective (i.e., man-machine mix). The superiority of the man-machine mix forecasts has prompted some observers to suggest that the extension of numerical predictions to six days can assist in improving present methods, especially when modified by experienced forecasters (Namias, 1968, pp. 438-70).

The National Meteorological Center data indicate that the accuracy of the center's pressure and precipitation forecasts has improved substantially from 1958 to 1975. From 1958 to 1972, the accuracy of forecasts of pressure had increased by 28 percent at sea level and by 22 percent for the mid-atmosphere. The center's skill in precipitation forecasting increased by approximately 13 percent from 1960 to 1971 (Brown and Fawcett, 1972, pp. 1175-77).

These data make clear the fact that numerical prediction methods have enabled the National Meteorological Center to achieve impressive improvements in the accuracy of its forecast guidance. However, forecast verification records of the National Weather Service suggest that improvements at the center have not produced comparable improvements in the public's daily temperature and precipitation forecasts. Precipitation predictions for 150 to 250 National Weather Service forecast stations round the United States improved by 5 percent between 1959 and 1972. Larger improvements in prediction accuracy have been recorded for individual cities. Over a thirty-year period the proportion of correct daily precipitation and temperature forecasts for Chicago has improved by about 10 percent (Cooley and Derovin, 1972).

Increases in the accuracy of the National Meteorological Center's forecast guidance do not necessarily guarantee increased skill in local predictions of temperature and precipitation. A recent study conducted in M.I.T.'s Department of Meteorology monitored the accuracy of daily temperature and precipitation forecasts made at M.I.T. No secular increase in forecasting skill was found for the 1966-72 period. This finding was especially surprising since the M.I.T. forecasters made use of the National

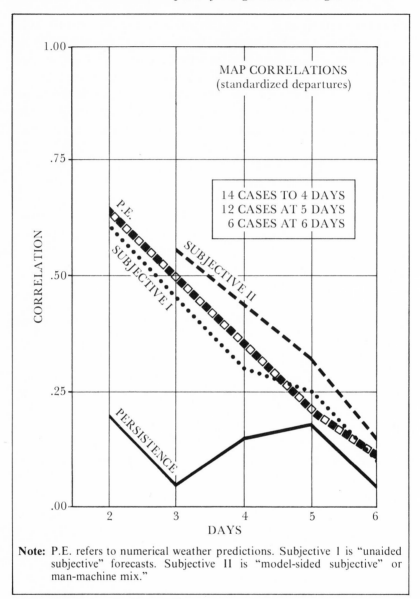

Note: P.E. refers to numerical weather predictions. Subjective 1 is "unaided subjective" forecasts. Subjective II is "model-sided subjective" or man-machine mix."

Source: J. Namais, "Long-Range Weather Forecasting: History, Current Status, and Outlook," *Bulletin of the American Meteorological Society* (1968), vol. 49, pp. 438-70, as reprinted in W. J. Maunder, *The Value of the Weather* (London: Methuen & Co., Ltd., 1970), p. 271.

Figure 4.1. Comparative Accuracy of Numerical Weather Predictions and Subjective Weather Forecasts

Meteorological Center's forecast guidance, which was steadily improving during this period. The paradox of more skilled center forecast guidance and no improvement in local Boston forecasts is explained in the limited state-of-the-art in adapting the center's regional forecasts to predict more localized weather conditions. Improvements in local weather forecasting may not depend on further advances in the center's forecast guidance; they may require, instead, an improved ability to cope with local weather influences (Sanders, 1973, pp. 1171-74).

When we turn from numerical prediction models to meteorological satellites, it becomes more difficult to assess the impact on forecasting accuracy. Until recently, the National Weather Service had not conducted any controlled tests comparing forecast accuracy with and without satellite data. As a result, there is no way to quantitatively estimate the effect of satellite data. Instead, it is necessary to rely on more subjective evidence.

Currently, satellites play a significant role in storm warning and forecasting. By the end of 1964, Tiros data had been used to discover and track more than a hundred hurricanes and typhoons. The National Environmental Satellite Service of NOAA estimates that since the inauguration of the operational satellite system in 1966, no tropical storm has gone undetected anywhere in the world. Similarly, the forecasting of severe local storms over the continental United States had advanced to the stage where area alerts and watches are issued routinely (Sanders, 1973, pp. 1171-74). The new geostationary satellites (SMS and GOES) now play a major role in this area (U.S. Department of Commerce, April 29, 1974, pp. 1-5; and Widger, 1966, p. 127).

The use of satellite data in forecasting parameters such as temperature, humidity, and wind is less extensive than in storm forecasting. Satellite data have been used as one input, but the forecasts of major parameters are primarily based on conventional methods of observation. Until 1969, the biggest difficulty in using satellite data was that forecasters could not develop a quantitative forecast of the wind or temperature on the basis of cloud pictures. The advent of vertical temperature soundings with Nimbus III proved that temperature and humidity data could be provided from satellites. While the National Meteorological Center has established an experimental program to use these data, uncertainty about the accuracy of satellite sounding data restricted their use before 1973. Routine use of satellite remote sounding data in daily public forecasts provided by the National Weather Service did not begin until mid-1973. It is still too early to accurately assess the impact of these satellite data on forecast accuracy (Burtsev and Cistajakov, 1971, pp. 19-36). This is borne out in a personal letter to the author from Dr. Duane S. Cooley, Technical Procedures Branch of the National Weather Service. In response to a question concerning the use of satellite data in daily weather forecasts, Dr. Cooley wrote on April 15, 1973:

> We are just beginning to develop quantitative information on the influence of satellite remote sounding data on operational forecasts. In the past, the use of satellite information in preparing initial analyses had largely been limited to location of storm centers over the oceans and extraction of approximate high and low level wind information from clouds over the oceans.

In addition to providing weather services to the general public, the National Weather Service furnishes a variety of specialized forecasts. Included among these specialized services are the Agricultural Weather Service, the Aviation Weather Service, and the Marine Weather Service. There are indications that the accuracy of these specialized weather forecasts has improved more than the accuracy of the forecasts disseminated to the general public. Although no quantitative trend data tracking the accuracy of specialized weather forecasts are available, the increased use of such services appears to reflect a growing recognition of their importance and increased user satisfaction with forecast accuracy. For example, the improved accuracy of marine forecasts has induced shipping companies to make greater use of weather routing — using specialized weather forecasts to gain all of the advantages of the prevailing weather conditions, but to avoid most of the disadvantages. In addition, private weather consultants are playing a growing and more seminal role in shipping (Maunder, 1970, pp. 117-19).

Economic Value of Weather Forecasts

Most human activities are either directly or indirectly influenced by weather. Yet until recently the value of more accurate weather forecasts was unexplored by both social scientists and meteorologists. The increasing recognition of the potential for an economic evaluation of weather forecasting is due to the costly technological advances described above. The public expenditures associated with these advances dwarf the relatively small costs of earlier weather services. As the investments in operational meteorology and atmospheric research have increased, greater attention has been focused on the prospective economic and social benefits of these investments.

In general, benefits can be realized from weather forecasts if the weather itself can be modified or controlled. Currently, weather modification is only feasible in limited areas for specialized purposes. Protecting activities and property is regarded as the most practicable of the alternatives, and economic studies have focused on the protection alternative.

The total value of annual weather-caused losses in the United States is huge. Thompson's survey (1972, pp. 7-11) of agricultural, industrial, and other activities suggests that the annual value of weather-caused losses was $12.7 billion. Approximately $5.3 billion of this total were losses which could have been averted if adequate warnings were provided. However, it is important to recognize that all of these "protectable losses" cannot be avoided since the costs of protection must be taken into consideration. Perfect weather forecasts can save about 15 percent of these protectable losses. While this is a relatively modest proportion of total protectable losses, the absolute value of these savings is very large — $739 million according to Thompson's estimate.

Achieving these economic benefits does not depend entirely on improving weather forecast accuracy. A significant portion of the economic benefits can be attained through proper use of present weather forecasts. The available evidence indicates that weather forecast users (and nonusers) do not fully exploit the value of current weather forecasts. This may be due to incomplete dissemination of forecast information or poor operating decisions. Thompson calculates that better use of existing forecasts accounts for 44 percent of his projected annual savings of $739 million (Table 4-2).

A striking proportion of the potential economic gains are in agriculture. Although it is generally acknowledged that agriculture is the single most important sector subject to the

Table 4-2

Potential Annual Savings Due to Improvements in
Weather Forecasting in the United States

(Millions of dollars)

| Activity | Improvements in | | Total Gains[a] |
	Use	Accuracy	
Agriculture	$250.3	$316.7	$567.0
Aviation (commercial)	1.4	2.2	3.6
Construction	13.1	18.4	31.5
Communications	0.3	0.4	0.6
Electric power	0.5	0.8	1.3
Energy (e.g., fossil) fuels	(b)	0.1	0.1
Manufacturing	8.1	11.9	20.0
Transportation (rail, highway, and water)	1.3	1.9	3.2
Other (general public, government, and so on)	47.3	64.5	111.8
TOTAL[a]	322.2	416.9	739.1

[a] Sums may not balance because of rounding.
[b] Less than $50,000.

Source: J. C. Thompson, "The Potential Economic Benefits of Improvements in Weather Forecasting" (San Jose: Department of Meteorology, California State University, September 1972), p. 53.

vagaries of the weather, the estimates of potential economic gains vary widely. There is no apparent way to reconcile quantitative estimates made for the total agricultural sector with case studies of individual agricultural crops. In 1972 Thompson estimated the potential savings in agriculture at $567 million; an earlier Stanford University study (1966) placed savings at $313 million. Yet a case study of Wisconsin hay farmers concluded that more accurate weather forecasts could produce savings in the range of $42 million to $84 million in Wisconsin alone, depending on the particular harvesting strategy employed. Similar differences exist between aggregate national estimates and case study projections for construction, manufacturing, recreation, and other sectors of the economy (Smith, Boness, and Smith, vol. II, pp. 148-68).

In certain sectors, such as agriculture, the predicted economic gains should be adjusted to take into account longer run consequences of improved weather prediction. In his study of the Oregon pea industry, Lee Anderson (1973, pp. 115-25) adjusted predicted economic gains downward because processing capacity limitations would prevent processors from accepting all the additional pea supply. Another illustration is L. B. Lave's raisin study (pp. 151-64), which suggests that $10 million worth of raisins could be saved each year through improved decision making. However, the improved decision making would lead to excess supply, thus forcing the less efficient raisin farmers out of business and making land available for other purposes. An estimated $20 million to $40 million of additional savings could be realized through proper use of this land.

In nonagricultural sectors, such as construction and trucking, institutional barriers may prevent organizations from instituting appropriate protective action. The large number of small firms and the high degree of labor intensity in these industries are two such constraints. A preponderance of small firms in construction implies a small number of projects and a localized geographical base of operations. Both factors would limit the firm's ability to adjust to weather forecasts. Furthermore, small size implies limited planning efforts, making it unlikely that the firm would be able to use weather information more than one or two days in advance.

Another constraint is the high wage structure in construction and transportation. This often means that it is less costly from the firm's standpoint to incur weather losses than it is to speed up work by employing additional workers or by paying premium rates for overtime work. Moreover, even if firms wished to speed up work, skilled craftworkers are not always readily available. Union-imposed limits on layover time for drivers in the transportation industry and restrictions on the type of job a particular worker may perform are additional examples of institutional barriers (Smith and Boness, vol. III, pp. 5-106).

Potential economic gains from accurate forecasts vary considerably according to the length of the forecast period. In all, the greatest economic benefits are obtained by improving longer range predictions. More than half of the annual savings estimated in Thompson's study are for forecast periods of thirty days or more (Table 4-3). Predictions for periods of different length vary in their importance to different sectors of the economy. Thus the high importance attached to the ninety-day forecasts in the table is primarily due to the value of such seasonal forecasts to agricultural users. In contrast, aviation users attach primary importance to exceedingly short-range predictions (i.e., three to twelve hours) and little to forecasts beyond five days (Smith and Boness, vol. III, pp. 5-106).

The most destructive weather phenomenon in economic terms is the tropical storm. Hurricanes have in fact caused more damage in the United States than any other type of catastrophe, with seventeen thousand lives lost and property damage of $5 billion in the first sixty years of the century. An example of the destructive consequences of hurricanes occured on June 27, 1957, when Hurricane Audrey moved inland over Louisiana. In less than two days, more than five hundred people lost their lives, forty thousand to fifty thousand cattle died, and property damage was estimated at between $150 million and $200 million (Thompson, 1972, pp. 54-55). The economic losses in the United States and Canada from selected hurricanes are shown in Table 4-4.

While the average annual damage from hurricanes in the United States is estimated to be $300 million, the cost of an individual hurricane has been as high as $3 billion (Hurricane Agnes in 1972). Even the best forecast cannot prevent the structural devastation brought about by a severe hurricane like Camille, Hazel, or Betsy. However, it is likely that 15 percent of residential and commercial property damage could be avoided if the entire population were to heed hurricane warnings and take action to protect property. With estimated annual property damage of $72 million, complete protection would save $10.8 million. Since only 20 percent of the population heeds warnings, $8.6 million is therefore lost unnecessarily (Anderson and Burnham, 1973, pp. 126-33; and Simpson and Hebert, 1973, pp. 323-33).

Hurricane forecasting has improved from virtually no warning during the first and second decades of this century to a current forecast 24-hour displacement error of about a hundred miles (Maunder, 1970, p. 11). This measure of forecast accuracy means that forecasters can predict the point of a hurricane landfall; i.e., where the hurricane first touches down inland within a

Table 4-3

Potential Annual Savings Due to Improvements in Weather
Forecasting in the United States, by
Forecast Period Length

(Millions of dollars)

| Forecast Period | Improvements in | | Total Gains[a] |
	Use	Accuracy	
One to five hours	$ 3.8	$ 7.6	$ 11.5
Six to eleven hours	8.1	14.1	22.2
Twelve to 36 hours	42.9	69.7	112.6
Two to five days	79.1	94.2	173.3
Thirty days	82.0	86.9	168.9
Ninety days	106.3	144.4	250.7
TOTAL[a]	322.2	416.9	739.1

[a] Sums may not balance because of rounding.

Source: J. C. Thompson, "The Potential Economic Benefits of Improvements in
Weather Forecasting" (San Jose: Department of Meteorology, California
State University, September 1972), p. 55.

hundred miles. Economic gains would result from improved
accuracy; if the present error were reduced by 50 percent, the
20 percent who now protect themselves and their property would
realize about $15.2 million instead of $10.8 million (Sugg, 1967,
pp. 143-46). The economic gains are dwarfed by the potential
benefits that would result if a higher proportion of the population
used protective measures. If the proportion of the population
taking such measures rose from 20 to 28 percent, the potential
saving would be $25 million; an increase to 68 percent would
yield a total saving of approximately $60 million (Anderson and
Burnham, 1973, p. 126).

The economic value of tornado forecasting has received less
attention than the economics of hurricanes. Two factors probably
account for the more limited investigation. First, early statistics
regarding the damage caused by tornadoes were less extensive than
those for hurricanes. Second, those statistics that were available
suggested that tornadoes, while costly, were far less destructive

Table 4-4
Hurricane Damage in the United States and Canada

Hurricane	Year	Damage
Agnes	1972	$3,100.0
Camille	1969	1,421.0
Betsy	1965	1,420.0
Diane	1955	800.0
Carol	1954	450.0
Carla	1961	400.0
Unnamed (in New England)	1938	389.0
Donna	1960	387.0
Hazel	1954	252.0
Dora	1964	250.0
Beulah	1967	200.0
Audrey	1957	150.0
Cleo	1964	129.0
Hilda	1964	125.0
Unnamed (in Florida)	1926	112.0
Isobell	1964	10.0
Alma	1966	10.0
Unnamed (in Florida Keys)	1966	5.0
Inez	1966	5.0
Ginny	1963	0.4[b]

[a] Millions of dollars.

[b] The loss was offset by the rains that proved beneficial.

Source: A. L. Sugg, "Economic Aspects of Hurricanes" (*Monthly Labor Review*, 1967), vol. 45, no. 3, pp. 143-46; A. L. Sugg and R. L. Carrodus, "Memorable Hurricanes of the United States since 1873" (*ESSA Technical Memorandum*, WBTM-SR-42, 1969); W. J. Maunder, *The Value of the Weather* (London: Methuen & Co., Ltd., 1970), p. 9; and R. H. Simpson and P. L. Herbert, "Atlantic Hurricane Season of 1972" (*Monthly Labor Review*, April 1973), vol. 101, no. 4, pp. 323-33.

than hurricanes. A mid-1950s study estimated total tornado damage for the United States during the 36-year period from 1916 to 1952 at about $15 million annually (Flora, 1956).

More recent studies indicate that losses due to tornadoes since 1960 are far greater than these earlier estimates. A study of the 1963-70 period suggests that losses from tornado destruction approaches an average of $200 million per year (Sanders, 1971, pp. 446-49). These tornado losses reflect the damage inflicted by the Palm Sunday storms in the lower Great Lakes on April 11, 1965, and the disasters at Topeka, Kansas, on June 8, 1966 (Galway, 1966, pp. 144-49) and at Lubbock, Texas, on May 11, 1970.

The widespread tornado outbreak of April 3 and 4, 1974, attests to the awesome destructive potential of tornadoes. This outbreak was a once-in-a-century event that far exceeded earlier tornadoes in terms of numbers, length of tracks, total area affected, deaths, and damage. Somewhere between 75 and 85 tornadoes occurred within the area encompassed by a line from Chicago southward almost to the Gulf of Mexico and eastward to the Appalachians. Property damage attributable to these storms has been placed at $540 million; in addition, 328 deaths and 6,142 injuries were recorded (U.S.Department of Commerce, April 29, 1974).

Conclusions

Four conclusions emerge from this review of the development of meteorological satellites, their role in weather forecasting, and the economic value of weather prediction.

Conclusion one: Meteorological satellites have transformed methods of weather observation; however, it is not yet possible to measure any specific improvements that may have occurred in the accuracy of temperature and precipitation forecasts disseminated to weather information users.

Since their inception in 1960, meteorological satellites have transformed methods of weather observation. The unique value of satellite data stems from several factors which distinguish these data from other kinds of environmental observations. One characteristic is paramount: the effectiveness of the satellite in providing information, routinely and dependably, in areas where conventional data are sparse or absent. Closing observational gaps over oceans and remote land areas is cited time and again as the basic contribution of space satellites to improved weather forecasts.

Another factor is the rapidity with which forecast offices receive pictures and weather charts transmitted directly from satellites. In some remote areas, satellite data are received and used hours ahead of the regular data arriving through normal communication channels. A third aspect of satellite data that has proved of great value in numerous applications is the overall view of the earth and atmospheric features obtained from orbital altitude. Observations from this vantage point reveal large-scale organization and structure of cloud systems that are not attainable from high-flying aircraft or conventional ground-based observations.

Despite the advantages of meteorological satellites, there are not yet quantitative data to suggest how satellite data have improved the accuracy of forecasts for the users of weather information. There is some evidence that satellite data have had a significant impact on the specialized forecasts for marine, aviation, and agricultural interests, but specific measures of accuracy improvement await the results of controlled numerical prediction experiments instituted in 1974.

Interest in the economics of weather prediction has increased in recent years. However, all of the studies reviewed in our analysis refer to the potential gains from improved forecasts. Despite a general belief in the value of more accurate forecasts, no actual economic gains or benefits have been identified in these studies. It is also significant that studies of potential gains tend to focus on the economic value of perfect weather forecasts. It is highly unlikely that such studies could be refined enough to estimate the economic value of intermediate improvements in forecast accuracy.

Conclusion two: To date, the major benefits derived from operational satellite systems result from improved storm warnings and forecasts.

Satellites now provide daily blanket coverage of the tropical oceans, permitting detection and surveillance of every disturbance.

Satellite data permit earlier and more dependable forecasts of hurricane development, future path, and intensity. Although no tropical storm has gone undetected since the installation of the operational satellite systems, no quantitative data exist on the improvement in hurricane forecast accuracy that has resulted.

The forecasting of tornadoes and severe local storms over the continental United States has also advanced to the stage where area alerts and watches are issued routinely. However, the science of tornado forecasting has not reached the stage of specifying the time and precise location at which a tornado will strike. Individual storms can be discovered only by personal sighting, or by weather radars with a range of a hundred to 120 miles. These methods may provide warning only a few minutes in advance, which may be too late. Geostationary satellites now provide pictures that cover areas beyond radar range, and show the early stage of storm development when distinctive cloud patterns first begin to form. These patterns appear in the satellite picture several hours before the clouds develop to a point where the water droplets in them are large enough to be detected by radar.

The annual cost of the damage produced by hurricanes, tornadoes, and severe storms in the United States is currently in the range of $400 million to $500 million. Assuming that satellite data have increased the accuracy of storm forecasts by 50 to 100 percent since 1966, the potential economic benefits through avoidance of property damage over the 1966-73 period approximates $100 million. However, because only 20 percent of the population protects during any storm, economic savings have probably been limited to a maximum of $20 million over this period. The true potential economic and social benefits of an accurate storm warning system will be realized only when there is a substantial increase in the proportion of the affected population that takes protective measures.

Conclusion three: The full potential of meteorological satellites in weather forecasting has not yet been realized. Advances in satellite meteorology should permit significant improvements in weather forecasting accuracy by 1980.

In assessing the fourteen years of satellite experience, one must recognize that satellites did not provide quantitative data on temperature and humidity until April 1969. Prior to 1969, satellite data were not available in a format that could be a direct input to mathematical models. Thus meteorologists have had less than five years' experience at coupling satellite data with numerical prediction models.

To date, quantitative measurements have been primarily successful in the optical and infrared ranges of the spectrum. While optical and infrared instrumentation continue to improve, clouds have proved a serious problem in infrared measurement of temperature. Consequently, the effective use of microwave sounders on Nimbus V, launched in 1973, to measure temperature down to the earth's surface, even in the presence of clouds, constituted a valuable milestone in satellite instrumentation development.

The operational meteorological system will be updated in 1978 with the addition of sophisticated new scanning instruments, including microwaves. The use of this improved satellite data in refined numerical prediction models is expected to yield marked improvements in one- to three-day weather forecasts, particularly in those coastal areas where weather comes from data-sparse oceanic regions.

Longer range weather forecasting accuracy is expected to improve considerably as a result of the Global Atmospheric Research Program. This program reflects the optimism of the world meteorological community regarding space observational capabilities. The first global experiment of the program (scheduled for 1978-79) will be an internationally supported research effort whose objective is to bring the entire global atmosphere under constant surveillance for a period of about a year. This is to be done with satellites, balloons, ground stations, buoys, and ships. The data collected by this observational network will be distributed over the global telecommunication system organized by the World Meteorological Organization to research and forecasting centers all over the world. These centers will use the data in numerical prediction models to determine how far the time range of effective prediction can be extended.

The data collected from the program's 1978-79 first global experiment will exceed by orders of magnitude anything available before. Prediction experiments with these data will indicate the observational requirements for long-range prediction models. Some observers envision that the data from this first experiment will permit weather prediction centers to issue weather forecasts for about a one-week period with the accuracy now available for one day (Tepper, 1974).

Conclusion four: Very large economic gains are realizable with present levels of weather forecast and storm warning accuracy.

It is unlikely that the contributions of meteorological satellites and numerical weather prediction in meteorology will be fully exploited until two major barriers are overcome. First, substantial improvements in the dissemination of weather information are required. The most pressing need in this respect is to provide the user with the specific type of weather information that he or she requires. As an illustration, most economic models used to estimate potential savings from better forecasts focus on how users should make optimum use of weather information in decision making. Such models presume that weather predictions include information concerning their uncertainty. It is only recently, however, that the National Weather Service began to meet this requirement by disseminating "probability forecasts." As of 1973, operational probability forecasts existed only for the occurrence of measurable precipitation (personal correspondence with Dr. Duane Cooley, 1973).

Second, decision making by farmers, businessmen, builders, and other users of weather information is far from optimal. The inadequacy of present decision strategies is demonstrated by Thompson's estimate that 44 percent of potential economic gains could be achieved through better use of current forecasts (Table 4-2). Furthermore, the economic benefits of more accurate weather forecasts are unlikely to materialize unless users employ decision strategies which capitalize on the improved information.

In light of these operational barriers, it is useful to examine the share of federal meteorological expenditures allocated for dealing with these problems. Of the estimated 1974 federal expenditures for meteorological operations, 15 percent is allocated to dissemination. However, the more significant ratio is the low 2.6 percent of meteorological research expenditures budgeted for dissemination research. Given the severity of the dissemination problems and the high economic payoff from solutions, dissemination research appears to be substantially underfunded.

In all, it is evident that the economic impact of NASA's meteorological satellite program cannot be reliably assessed fifteen years after the launching of the first weather satellite. While this is partly due to the problems of inadequate data, it is the case primarily because the true impact of such a program can be measured only over a much longer time range. It will be at least 1980 before the full potential of satellites in weather forecasting begins to be realized; and even then, the limitations on dissemination and utilization of available forecasts may remain open issues.

References

Anderson, Lee G. "The Economics of Extended-Term Weather Forecasting." *Monthly Weather Review* (February 1973), vol. 100, no. 2.

_____; and Burnham, John M. "Application of Economic Analyses of Hurricane Warnings to Residential and Retail Activities in the U.S. Gulf of Mexico Coastal Region." *Monthly Weather Review* (February 1973), vol. 100, no. 2.

Brown, H.; and Fawcett, E. "Use of Numerical Guidance at the National Weather Service's National Meteorological Center." *Journal of Applied Meteorology* (November 1972), vol. 54, no. 11.

Burtsev, A. I.; and Cistajakov, A. D. "The Use of Satellite Information for Weather Analysis and Forecast." In *World Meteorology*. Geneva, Switzerland. 1971.

Cooley, Duane S.; and Derovin, Robert G. "Long-Term Verification Trends of Forecasts by the National Weather Service." NOAA Technical Memorandum NWS FCST-18. Silver Spring, Maryland: U.S. Department of Commerce. May 1972.

Delbeq, André; and Filley, Alan. "A Study of the Weather Satellite Program Management and Organization System." In *Multidisciplinary Studies of the Social, Economic, and Political Impact Resulting from Recent Advances in Satellite Meteorology*. Volume IV. Madison: University of Wisconsin.

Flora, S. D. *Tornadoes of the United States.* Norman: University of Oklahoma Press. 1953.

Galway, J. G. "The Topeka Tornado of June 8, 1966." *Weatherwise* (1966), vol. 19, no. 4.

Hubert, L.; and Lehr, P. *Weather Satellite.* Waltham, Massachusetts: Blaisdell Publishing Co. 1967.

Hughes, Patrick. *A Century of Weather Service.* New York: Gordon and Breach. 1970.

Lane, F. W. *The Elements Rage.* Philadelphia, Pennsylvania: Chilton Books. 1965.

Lave, L. B. "The Value of Better Weather Information for the Raisin Industry." *Econometrica* (undated), vol. 31, nos. 1 and 2.

Legislative Reference Service, Library of Congress. *Meteorological Satellites.* Washington, D.C.: U.S. Government Printing Office. 1962.

Maunder, W. J. *The Value of the Weather.* London: Methuen & Co., Ltd. 1970.

Namias, J. "Long-Range Weather Forecasting: History, Current Status, and Outlook." *Bulletin of the American Meteorological Society* (1968), vol. 44.

National Aeronautics and Space Administration. *Aeronautics and Space Report of the President, 1974 Activities.* Washington, D.C.: National Aeronautics and Space Administration. 1975.

Popkin, Roy. *The Environmental Science Service Administration.* New York: Frederick A. Praeger, Inc. 1967.

Sanders, F. "Skill in Forecasting Daily Temperature and Precipitation: Some Experimental Results." *Bulletin of the American Meteorological Society* (November 1973), vol. 54, no. 11.

_____. "Toward Defining Human Needs: How Does the Atmosphere Hurt Us?" *Bulletin of the American Meteorological Society* (June 1971), vol. 52, no. 6.

Shuman, Frederick G. "Numerical Weather Prediction Capabilities in the 70's." Washington, D.C.: National Meteorological Center. March 1972. Unpublished paper.

_____. "The Research and Development Program at the National Meteorological Center." Washington, D.C.: National Meteorological Center. March 1972. Unpublished paper.

Simpson, R. H.; and Hebert, Paul J. "Atlantic Hurricane Season of 1972." *Monthly Weather Review* (April 1973), vol. 101, no. 4.

Smith, K.; and Boness, F. "The Economic Impact of Recent Advances in Weather Satellite Meteorology." In *Multidisciplinary Studies of the Social, Economic, and Political Impact Resulting from Recent Advances in Satellite Meteorology.* Volume III. Madison: University of Wisconsin.

_____; Boness, F.; and Smith, D. "The Impact on the Hay Industry of Improved Satellite Weather Forecasts." In *Multidisciplinary Studies of the Social, Economic, and Political Impact Resulting from Recent Advances in Satellite Meteorology.* Volume II. Madison: University of Wisconsin.

Space Science and Engineering Center. *Multidisciplinary Studies of the Social, Economic, and Political Impact Resulting from Recent Advances in Satellite Meteorology.* Volumes I, II, and III. Madison: University of Wisconsin, Space Science and Engineering Center.

Stanford School of Engineering. "Stanford Proposal for an International Network for Meteorological Analysis and Prediction." Palo Alto, California: Stanford University. 1966.

Sugg, Arnold L. "Economic Aspects of Hurricanes." *Monthly Weather Review* (March 1967), vol. 45, no. 3.

Tepper, M. "On the Contribution of Space Observation to the Future of Meteorology." Paper presented to UNESCO Conference on the Impact of Science on Society. March 6, 1974.

Thompson, J. C. "The Potential Economic Benefits of Improvements in Weather Forecasting." San Jose: California State University, Department of Meteorology. September 1972.

U.S. Department of Commerce. *The Federal Plan for Meteorological Services and Supporting Research, Fiscal Year 1974.* Washington, D.C.: U.S. Government Printing Office. 1974.

—————. *First Five Years of the Environmental Satellite Program: An Assessment.* Washington, D.C.: U.S. Government Printing Office. February 1971.

—————, National Oceanic and Atmospheric Administration. "The Widespread Tornado Outbreak of April 3-4, 1974." Washington, D.C.: U.S. Department of Commerce, April 29, 1974. Preliminary report.

Widger, William K., Jr. *Meteorological Satellites.* New York: Holt, Rinehart, and Winston, Inc. 1966.

Chapter Five

●

From Case Studies to Generalizations

The case studies of the semiconductor and computer industries, the field of astronomy, and the gains from meteorological satellites have each provided interesting findings regarding the impact of NASA on particular institutions. But they also yield another benefit. Together with the previous studies commissioned by NASA, these three case studies provide a basis for formulating generalizations regarding the impact of large-scale public programs. The generalizations relate to four significant issues: the determinants of program impacts, the relationship of impacts to program evaluation, the link between program activites and technological change, and the manpower implications of program impacts.

Determinants of Impacts

The first set of generalizations suggested by our studies is related to the question: How do impacts come about? The relationship between budgetary and programmatic decisions made by administrators and the impacts generated by their programs is not random. Impacts do not merely happen; they are not due to serendipity. Goals, budgets, programs, and operating decisions established and carried out by managers of large public programs shape these programs' impacts.

A systems view of the major stages of planning and decision making which shape a public program is helpful in analyzing the determinants of program impacts. The four key stages are (1) the formulation of a social goal (or goals), (2) the establishment of an agency budget, (3) the development of legislative and administrative programs, and (4) operational decision making. The process is shown in Figure 5.1.

The first phase of public program planning is the establishment of goals. Two significant characteristics of a program goal are mission priority and mission specificity. Mission priority reflects the degree of national support for the social goal. If there is substantial national support for the objective, and it is not perceived as impinging unfavorably on other valued social goals, high mission priority is likely. Mission specificity reflects the precision of desired program achievements. How specific are the time, quantity, and quality dimensions of goals? Is the mission specified as "placing a man on the moon by the end of the decade," or is it a more broadly worded objective such as "improving the quality of health care for the aged"? Both mission priority and mission specificity are important influences on subsequent budgetary and programmatic decisions.

Our studies indicate that budgetary decisions are the single most important determinants of program impacts. However, *absolute investment* is not the key factor; the primary determinant is the *size of the program's expenditures* relative to the *existing budget* of the institution affected. The impact of a public program upon a given institution — an industry, a science, a community — is primarily determined by the relative rate of public investment in that institution. Thus in a sector whose total expenditures are $10 billion to $20 billion, a public program with a $500 million budget will not generate significant program impacts; but in a sector with total expenditures of only $5 billion, a public program with the same $500 million budget may produce substantial impacts.

The importance of an agency's budget in influencing institutions was a common theme that emerged from the case studies. Consider the impact of the space program on astronomy: NASA's expenditures on astronomy have not been huge; they are small compared to total space expenditures or, for that matter, when compared to NASA's spending on computers. However, NASA's financial support relative to the budgets of existing astronomical institutions has been large, and the effects have been correspond-

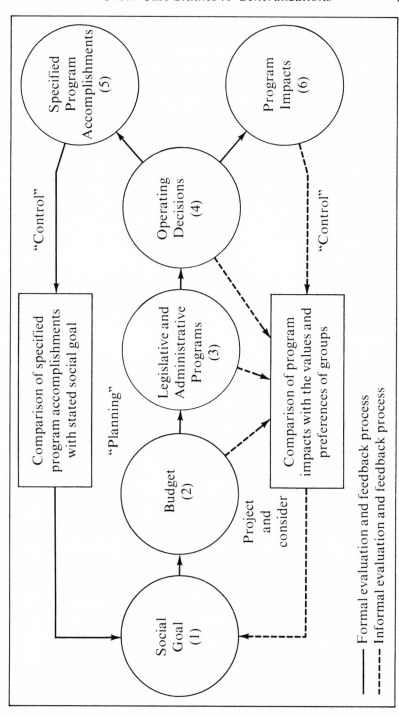

Figure 5.1. Public Program Impacts — Determinants and Evaluation Processes

Social Goal (1)

Budget (2)

Legislative and Administrative Programs (3)

Operating Decisions (4)

Specified Program Accomplishments (5)

Program Impacts (6)

Comparison of specified program accomplishments with stated social goal

Comparison of program impacts with the values and preferences of groups

"Control"

"Control"

"Planning"

Project and consider

—— Formal evaluation and feedback process

----- Informal evaluation and feedback process

ingly large. Thus NASA was able to stimulate the study of previously moribund areas, provide new observational facilities and instrumentation techniques, and attract new manpower to the field through the space program's considerable financial leverage.

A public program's social goal is elaborated through the development of legislative and administrative programs. In selecting legislative and administrative strategies, managers make two significant sets of allocation decisions. First, there is the allocation of the budget among major program elements. These decisions fix the size and duration of programs, determine the role of outside vendors and contractors, and shape the involvement of university and other nonprofit institutions. For example, NASA's initial investment in astronomy was resolved during this phase. A second set of allocation decisions relates to the division of program funds among regions. Decisions to construct or locate new facilities in particular geographic areas determine the regional socioeconomic impacts of the public program. The technological requirements of NASA's manned space flight dictated certain locational constraints which favored communities in the "southern crescent." The economies and social structure of these communities were substantially modified by the space program's presence.

The two broad classes of allocation decisions made during the development of legislative and administrative strategies produce the resources available to and constraints imposed upon program managers as they make and implement operating decisions. Program budgets, the number and location of facilities, the size and skills of the civil service and industrial work forces, the capabilities of university researchers, and similar considerations all shape the nature of decisions at the operating levels of public programs. In turn, these operating decisions directly yield specific accomplishments and program impacts. Purchasing decisions are one class of operating decisions that is a strong influence on program impacts. Does the program purchase conventional "off the shelf" equipment or does it contract for the development of new technology? Must contractors meet demanding technical specifications? The purchasing decisions NASA made reflected the space program's emphasis on new miniaturization technology and totally reliable computer systems. This emphasis proved important in accelerating technological advance in both the semiconductor and computer industries.

Evaluation of Impacts

The relationship between the planning and decision-making phases of a public program, the two types of program outputs, and evaluation processes is shown in Figure 5.1. A distinction is made between formal and informal program evaluations. Formal program evaluations compare specific program accomplishments with stated program objectives. In contrast, the informal evaluations compare program impacts with the value preferences of various groups. Although the informal evaluations are often less structured and less visible than formal evaluations, they are a significant factor in review of public programs.

Informal evaluation is further compounded by the absence of a consistent standard for measuring impacts. Different groups in society disagree over the goals and purposes of a public program and consequently judge the same program impacts against different standards. Moreover, the goals of various segments of society change, and thus the basis for assessing program impacts shifts. This dynamic aspect of impact assessment has especially affected NASA.

Although there was no direct public involvement in the formulation of the nation's space objectives, there was a high mission priority for space at the outset of the program. This high mission priority was established largely on the basis of appeals to national security needs. Furthermore, space goals did not appear immediately to impinge on other social goals; there were no lobbies opposed to space in the early years. Throughout, NASA was perceived as "benefiting all Americans." But in the process of achieving its mission to place a man on the moon, the huge space budget and manpower demands produced certain negative responses. The large-scale budget made space a highly visible target for groups promoting other social goals. As the Vietnam War claimed federal resources sought for Great Society programs, the defense-linked space program was viewed by many as a direct threat to domestic programs.

Another sector which raised objections to the space program was the scientific community. It criticized the manpower distortions created by NASA and the heavy engineering orientation of the program. These contradictory evaluations which characterized the space program are inevitable for all large public programs. Only the political process, where debate and compromise yield at least a tentative consensus among the major groups, can resolve these differences.

Technological Dimensions

The role of large public programs in advancing science and technology is paradoxical. The managers of a large government program face a major dilemma in making scientific and technical decisions. If they are to ensure the accomplishment of specified program objectives, they must plan to use only currently available scientific knowledge and technology. They must be technically conservative and focus on exploiting the existing state-of-the-art rather than carrying out revolutionary scientific advances within their program. Consequently, it is unlikely that public programs will originate major innovations, although they can play an important role in technological advance. Given the constraints imposed by specified program objectives, it is unrealistic to expect evaluations to show that a government program deserves sole or direct credit for generating important scientific developments.

While public programs usually concentrate on exploiting the existing technology, they do not necessarily receive a "free ride" from other institutions, such as industrial firms or universities. Programs such as space and defense have increased the speed at which the state-of-the-art advances. The case studies uncovered several examples of the space program's acceleration influences. In astronomy, NASA accelerated the changes that had already been initiated by nonspace-related advances in radio astronomy. While NASA did not play the role of "pioneer" in the computer and semiconductor industries, it did contribute to accelerate technological trends toward extensive miniaturization and high reliability. Commercial products such as pocket calculators and digital watches, promoted as space-age innovations (complete with "Apollo" trademark), did not result from the space program, but they probably have become available to consumers much earlier because of NASA.

Public programs also play a weighty role in overcoming the resistance to technological change which inevitably limits early acceptance of new products and managerial methods. In the computer industry the "demonstration effect" provided by the space program's computers greatly increased the acceptance of the computer by the business world. The widely publicized use of computers for NASA space missions made business executives more willing to use computers for their industrial applications. The "demonstration effect" in computers is analogous to the "revelation effect" observed in astronomy. Space exploration led to the unexpected discovery of new objects and phenomena in the solar system, the galaxy, and beyond. These revelations caused astronomers to alter long-held traditional views of the universe. Eventually, the discoveries led to the development of new astronomical specialties, such as particles and fields, and X-ray astronomy.

Another characteristic of technological impacts is that they rarely yield quick economic payoffs. The process of converting a significant technological advance into new products or processes is complex and time consuming. Studies of major twentieth century innovations indicate that the time interval between an invention and its commercial use averages between ten and twenty years. Although such studies indicate that the average time interval between invention and innovation for government-supported innovations is shorter than the average for privately supported innovations, the lag is still substantial (Lynn, 1966). Consequently, we should not expect the significant technological impacts of public programs to produce economic returns in the short run.

The meteorological satellite is a good example of a government-supported technological advance with a substantial time lag before economic returns are realized. The first weather satellite, Tiros I, was flown in April 1960, and since then satellites have transformed methods of weather observation. However, economic payoffs have not yet been realized from this important advance. In this case, a major limitation is that weather observation represents but one part of the total system which produces weather forecasts. The improved observational data provided by the space satellites must first be integrated into the numerical prediction models used for weather analysis and forecasting. Then dissemination methods for presenting weather information to the public must be altered so that the public receives forecasts in a useful format. These complex integration problems must be solved before the full economic benefits of meteorological satellites are realized.

Manpower Dimensions

Important manpower impacts have occurred in each institution related to the space program. Earlier NASA impact studies documented the notable influence of the manned space flight program on employment growth and economic development in communities comprising the "southern crescent" along the Gulf of Mexico. The influx of scientific, technical, and professional personnel to smaller communities produced marked improvements in the educational systems at the elementary, secondary, and college levels.

By 1966 the space program employed 5.5 percent of the nation's engineers and scientists. Because NASA officials were concerned that the agency's demands for skilled scientists and engineers would create shortages in other sectors, it established a Sustaining University Program to award doctoral grants and pre-doctoral traineeships in a number of scientific fields. The astronomy case study provided some insight into the magnitude of these educational efforts. The space program's educational programs and project grants supported almost 40 percent of the 66 astronomy doctorates awarded in 1966-67. Funding by NASA, the National Science Foundation, and the Department of Defense, together with the new opportunities for experimentation provided by space exploration, increased the total number of astronomers threefold during the 1960-70 decade.

There was, however, a major flaw in the rationale for NASA's efforts to develop scientific manpower. Space officials erroneously assumed that the university is the only producer of manpower while the other two institutional members of the space program — industry and government — only consume manpower. And NASA underestimated the flexibility of a well-trained, highly skilled work force. Scientists shift among specialties with greater ease than is generally recognized. It is incorrect to assume that an employer only consumes manpower. The training provided for scientists, engineers, and managers by working on complex, advanced systems is as important a factor in increasing productivity as is schooling. High-technology government programs actually afford valuable developmental opportunities.

This training role for the space program was evident in the computer industry where NASA projects were a significant training ground for project engineers and managers. Computer firms have used this training by transferring substantial numbers of project managers and engineers from space-defense divisions to the commercial side of the business. In the case of astronomy, the new research opportunities induced a substantial number of physicists to shift into astronomy. By 1971 the number of researchers in astronomy with doctorates in physics exceeded those with earned doctorates in astronomy. Additional Ph.D. researchers left specialties, such as chemistry, engineering, and geology, to become astronomers.

The manpower development programs of NASA helped produce a supply of scientists which now exceeds current demand. In astronomy, for example, space program cutbacks and other declines in government funding have sharply reduced the need for astronomers. In retrospect, it is clear that NASA did not have to support Ph.D. students to the extent that it did. However, because of its substantial funding of astronomy and its specific efforts to attract and educate more astronomers, NASA has taken on a difficult responsibility. The instability of government funding and sudden changes in program plans have direct consequences for the affected astronomers.

There are those who contend that NASA's recent cancellation of astronomical programs constitutes a serious breach of trust, which has had destructive effects on the astronomical community. The message for public program managers is clear: Large public programs cannot become deeply involved in influencing manpower flows for the purposes of accomplishing specific agency objectives and avoid responsibility for subsequent negative impacts which may result. Indeed NASA's experience in astronomy suggests that government managers must be more sensitive to the longer term manpower consequences of their programs.

Reference

Lynn, F. "An Investigation of the Rate of Development and Diffusion of Technology in Our Modern Industrial Society." In *Report of the National Commission on Technology, Automation, and Economic Progress.* Washington, D.C.: U.S. Government Printing Office. 1966.

Chapter Six

●

Managing Program Impacts

Our study of the impacts of NASA carries significant implications for the overall management and strategic planning of large public programs. The long-range effectiveness — even survival — of such programs depends on their managers' abiltiy to shape the programs' impacts and to meet the shifting needs of a wide range of groups as well as to achieve specific objectives. Simply stated, the strategic management of impacts — economic, technological, and manpower — is essential to the success and survival of public agencies.

The magnitude of difficulty of the task does not, nevertheless make it any less important. Peter Drucker (1974, p. 18) states the case for impact management at the outset of his book:

> The manager has to be a craftsman. His first duty is indeed to make his institution perform the mission and purpose for the sake of which it exists — whether this be goods and services, learning, or patient care. But this is not enough. Any institution exists for the sake of society and within a community. It therefore has to have impacts; and one is responsible for one's impacts. In the society of institutions of the developed countries, the leadership groups, that is, the managers of the various institutions, also have to take social responsibility, have to think through the values, the beliefs, the commitments of their

153

society, and have to assume leadership responsibility beyond the discharge of the specific and limited mission of their institutions. This responsibility creates a major new challenge — and raises the most difficult problems, both of management and of political theory and practice. But it has become a fact.

In recent years, there has been growing recognition at the federal level of the need for predicting and controlling the impact of technology. In 1974 the congressional Office of Technology Assessment was established to assess the technological impacts of programs for which Congress appropriates money. Similarly, the Office of National R&D Assessment in the National Science Foundation has a major program element — socioeconomic effects of technological innovation — whose primary objective is "to identify, measure, and examine the effects of technological innovation" (National Science Foundation, 1974, p. B-1).

This chapter maps out a structured approach to strategic planning and evaluation which will aid the managers of NASA and other public agencies to respond to the challenge of impact management. Before outlining the approach, we must make explicit the key assumptions which were derived from our impact studies and which are the basis for the recommendations. In addition, the three phases involved in developing an overall NASA strategy should be examined. These three phases are:

(1) Impact group analyses

(2) The forecasting of program impacts

(3) Impact evaluation and control

Once these three stages have been examined, it will be possible to present a structured process for the management of impacts by NASA.

Key Assumptions and Managerial Implications

The findings of the case studies indicate that the following set of observations and assumptions provides a firm foundation for building a management strategy. First, NASA is a continuing organization whose activities, at any given time, consist of a bundle of programs. Each of the programs has a relatively independent existence in the sense that each ranges over a different time span, pursues a unique mission and objective, and is managed as a relatively autonomous enterprise. That such diverse programs could be comfortably accommodated under a single NASA umbrella is a tribute to the agency's senior administrators who have been able to conceive its mission in much broader terms than "getting a man to the moon before the Russians do."

The National Aeronautics and Space Administration, in common with other complex public, nonprofit organizations, continuously faces a challenge in executing its overall objectives and missions. Under the supervision and direction of Congress and the Executive branch, it is charged with advancing the "public good" and with serving the interests of "all Americans." Desirable and laudable as these goals may be, they become rather fuzzy in application. In reality, "the public" consists of a large variety of groups, classifiable along many axes: social, professional, geographic, or functional. Some of these groups may become deeply involved in, or deeply affected by, any specific program. Others may be almost totally distinterested or unaffected. Some groups consist of direct contributors to a given NASA effort, some may be direct beneficiaries of its efforts, while still others may be drastically affected by the methods chosen for their implementation.

Even relatively specific missions, such as landing a man on the moon, involve a variety of societal segments whose interests need to be considered and balanced. Site selections may involve competing geographic groupings. Technology decisions often affect different industry and professional sectors, while such conflicting demands as safety, low cost, deadlines, and career advancement are always present in any series of program decisions. As overall missions become broader in nature, it becomes more difficult to identify the groups most directly concerned with or impacted by program implementation. Thus a realistic approach to pursuing the "public good" or the "all-American interest" is to identify the major "impact groups" (a term that will be used for the purpose of brevity) involved in the bundle of NASA programs and to attempt to optimize their benefits. This is no mean task, particularly as the organization finds itself with limited resources and growing pressures from both Congress and the Executive branch.

In the long run, NASA's effectiveness is linked to its ability to fulfill the expectations and needs of a diverse clientelle, represented by major impact groups. These expectations are fulfilled (or frustrated) by program impacts as well as by specific program accomplishments. In some cases it is quite possible for the former to overshadow the importance of the latter. Program impacts, as defined in the preceding chapter, must of necessity be included in any assessment of the agency's total contribution.

In its earliest years, NASA's *specific* accomplishments were both popular and dramatic enough to satisfy, even enthusiastically fulfill, the expectations of key impact groups. It is not surprising, therefore, that its program impacts were treated as by-products which could be assessed *post facto*. The many impact studies, both voluntary and commissioned, followed this approach. The attitude was essentially: We've carried out our program, have competently executed its stated objectives; now let's get some experts to look back and assess the program impacts. Whom did they affect? Was it favorably? And, if possible, at what dollar cost or benefit?

The pattern of earlier years is changing rather dramatically and will continue to change. The relatively clear, dramatic program missions have been completed, NASA's impacts have broadened, and new "competitors" have emerged for meeting comparable expectations. Current and future impact groups will seek more diffused goals than any single program's specific accomplishments can meet. As a result they will look also to the impacts of each program to fulfill their needs. This is another way of saying that in the future, program impacts need to be carefully considered in the early planning stages of each program, rather than left to chance or for later researchers to discover.

The dominant managerial implication which emerges from these fundamental assumptions is the urgent need to formalize a process for the planning and evaluation of impacts. This requires an organized approach which (1) anticipates program impacts, as well as produces specific accomplishments, and (2) periodically assesses the total mix of impacts resulting from the aggregate portfolio of programs.

A second managerial implication derived from our observations and assumptions is that NASA must incorporate impact group analysis as an integral part of a formalized impact planning effort. That is, NASA must explicitly identify the groups concerned with individual space programs and describe their stakes, interests, and expectations. The case studies highlight one striking feature: the multiplicity and diversity of institutions which are impacted. The number of institutions modified and transformed by NASA is large. This process is not all one-way; the shifting interests, beliefs, and values of constituencies exert significant influences on the future direction of the space program.

The space administration's traditional approach to the assessment of its impacts on any segment of society has been essentially a *post-facto* audit or review. A more effective approach, which is consistent with the planning and evaluation framework essential for effective management, is to conduct an ongoing, periodic review of program impacts. Consequently, the third management implication of our case studies is that NASA must develop a system for impact control as well as for impact planning. Specifically, NASA requires an organized process for measuring and evaluating its impacts.

Impact Group Analysis

As noted in the previous section, NASA programs affect a large and diverse set of institutions. This set is dynamic in two significant ways. First, the value of any given group may vary considerably among NASA programs. University-based scientists, for example, may play a key role and therefore hold immense stakes in one NASA program and almost none whatsoever in another. Second, each group's concerns tend to shift (often dramatically) over time or across events. The clear implication for long-term strategy is twofold: A high level of awareness of the needs and values of contributors and beneficiaries in NASA activities and a sensitivity to changes in those needs is essential for performance which truly serves the "public good."

It is interesting to note that even private, profit-making organizations (whose objectives and missions are considerably less complex) find it necessary to analyze the impacts of their operations on a much wider set of impact groups than the simple "profit maximization" adage would suggest. The approach of at least one large progressive business corporation provides some guidelines to an intelligent attack on this problem. First, the corporation identified the full range of its impact groups ("stakeholders" was the term), beyond the tradition-bound

Big Three: stockholders, customers, and employees. It divided some categories, such as employees, into smaller, more homogeneous grouping (e.g., managers, professional, and nonexempt). It added several which were not previously considered — the local communities in which unit plants were located, for example.

Once the key stakeholders were identified, a brief list of their expectations was compiled. This was typically accomplished by asking business unit managers to answer, for each impact group, such questions as: What are the group's basic expectations about your business and performance ? Which needs has your business been satisfying? The answers were combined and classified into a series of statements about the *current* expectations and needs of key stakeholders. The next move was to project the most probable future trends. A variety of techniques, ranging from attitude surveys to scenario projections, was used in answering such questions as: Do we detect a shift in impact groups' needs or values? Are they likely to expect more or less of us? In what directions? How rapidly and strongly are such changes likely to occur? Significantly, in directing its managers' attention to stakeholder analysis, the corporation strongly emphasized concern with nonfinancial, nonquantifiable, and often non-economic needs and expectations. The use made of these stake-holder analyses, in formulating both unit and corporate strategies, will be seen later in this discussion. At this stage it is more useful simply to suggest an adaptation of this approach to NASA operations.

Specifically, the suggestion is for NASA to structure and formalize important planning activities which are doubtless already being informally or sporadically performed. A careful analysis of NASA's key impact groups should be undertaken. Considerable thought should be given to subdividing groups into relatively fine, homogeneous categories and to going beyond identifying only the most obvious ones. Some preliminary examples include: Congress and congressional committees; key departments of the Executive branch; employees of NASA — perhaps subdivided into such categories as managerial, scientific, engineering, permanent and on-loan; vendors and contractors; communities where NASA installations contribute significantly to the local economy; the academic community; and, finally, the general public (classified into a few relatively homogeneous categories). No outside consultant can have the intimate familiarity, the long experience, or the requisite sensitivity to analyze all of the relevant impact groups. This is a task which is best entrusted to NASA insiders with broad knowledge of both operations and political history.

Once the groups are identified, their stakes, interests, and expectations of NASA should be described. Business planning often classifies stakeholder interests and expectations into four major impact areas: economic, technological, political, and social. A similar classification would help public agencies identify and project likely impacts in the planning stage. Another useful device which might be adapted is the identification, at this stage, of other agencies (public or private) which also contribute to meeting the expectations of each of the major impact groups. This permits subsequent decisions in the selection of specific programs to consider NASA's unique strengths or comparative advantage in providing benefits most effectively and to identify programs which might rightfully be left for other agencies.

The next step involves the assessment of trends and probable future direction of impact groups' expectations and values. This is a complex effort in which social science researchers and astute political observers would probably prove helpful. The questions to be answered here are similar to those cited in the corporate analysis analogy: Are the groups' values, expectations, and dependence on NASA changing? In what direction and how rapidly? What are likely to be their highest level needs and expectations over the next several years? And who else is likely to supply these effectively?

In the case of NASA, where a complex set of both contributors and beneficiaries is likely to emerge, an additional refinement not often practiced in corporate planning would be appropriate: NASA should identify key spokesmen for various impact groups and use them as "listening posts" or "barometers." Periodic updating of group values and the testing of probable alternate reactions could then be performed relatively quickly and easily. As planning alternatives are being considered, a variety of "what if. . ." questions, as well as periodic survey questionnaires, could be directed to these carefully selected representatives.

Forecasting Program Impacts

The nature of NASA's mission and operations compounds the difficulties associated with forecasting program impacts. The planners of NASA have an especially complex task because of the space program's unique planning structure.

Management literature often distinguishes between two basic types of planning structures: standing plan structures and single-use plan structures. The former guide and direct the continuing, repetitive aspects of an operation. Standing plans, at high levels, establish policies for the repetitive aspects of purchasing raw materials, hiring personnel, or the quality levels of different product lines. At lower, more operational levels, standing plans specify which forms are to be filed where, which bolts are to be tightened in what order, and how many units are to be measured in each production batch. These are the plans which direct the day-by-day repetitive pattern of activities which characterize the operations of most business units. It is not surprising, therefore, that they constitute the bulk of the volume of plans and planning in large organizations.

Single-use plans, on the other hand, deal with one-of-a-kind nonrepetitive efforts of the business. In a typical organization, single-use plans would be used in the construction of a new facility or the implementation of a new organization structure. These are comparatively rare sets of planning activities which do not reoccur on a periodic basis. In such situations, single-use plans are the auxiliary or secondary set of activities designed to support the primary set of standing plans which are the crucial ones in achieving efficiency in performance.

The space administration's planning structure is an almost direct reversal of the conventional pattern. Single-use plans are the backbone of each program and mission, many of which are one-of-a-kind events in themselves. Very little repetitive activity is involved in NASA activities (particularly if the actual manufacturing operations in vendor's facilities are excluded from consideration). Standing plans in NASA focus on such activities as bookkeeping, office procedures, or personnel policies. In this sense, standing plans can be viewed as the secondary plans supporting a primary set of single-use plans which direct each program's progress toward its targeted objectives.

When single-use plans comprise the primary planning effort in an organization, the forecasting of impacts inevitably requires dealing with relatively high levels of uncertainty. With a limited repertory of directly comparable earlier programs, the forecasting of "single-point" values or relatively certain results becomes much less realistic and much more dangerous. The realities of a stochastic, or contingency, universe, rather than the simple-minded but comfortable assumption of a relatively mechanical and certain world, must be faced squarely. All of this becomes even more critical as one attempts to project or forecast impacts, as well as primary, accomplishments.

Forecasting the impacts of single-use plans is further complicated by the limited application that trial and error methods have for single-use plans. In improving and refining repetitive operations, trial and error (or iteration) plays a significant if not a dominant role. The very act of repetition provides the testing ground for the improved idea or the newly available material. One-of-a-kind activities, on the other hand, can only draw on trial and error by analogy or some similarity to earlier activities. Thus the more pioneering the effort, the less likely the contribution of trial and error methods. Instead of iterative methods, planning activities must be based on the ability to project and forecast the specific and unique results that may be anticipated from alternative courses of action.

Given the high orders of uncertainty associated with NASA's single-use plans, the expanded use of forecasting methodologies designed to cope with highly unpredictable futures is most appropriate. The probable magnitude of program impacts may be projected by a variety of techniques. These range from relatively simple single-point or three-point estimating to rather complex applications of risk analysis techniques, from almost wholly qualitative approaches, such as scenario writing or delphi-method forecasting, to the highly quantitative manipulation of simulation models. Estimates of "optimistic," "pessimistic," and "most likely" results can be made not only in terms of time dimensions (as is commonly done in PERT planning, from which this approach and terminology are borrowed) but also in terms of severity and direction of impact. In other instances it might prove more effective to canvass the opinions of a group of experts in the field, and by a Delphi-like exchange of forecasts and assumptions, reduce the range or probable impacts to some manageable set. On the other hand, a single expert or a research team might be requested to present a set of alternative future scenarios, with some indication of the linkages in their probabilities and some early warning indicators.

Impact Evaluation and Control

For managerial purposes evaluation and control have distinct meanings. Control is not seen as an accounting check on the use of assets or the confirmation of the accuracy of inventory reporting. Rather, it refers to that stage of the management process which asks the questions: Are we getting there? Will we meet our targets? Do we need to take any corrective action? Managerial control, therefore, emphasizes the need for the following components of a control system: (1) a target, benchmark, or "standard," against which progress must be measured; (2) a means of measuring — preferably in quantitative terms — the gap between the desired "standard" and the actual performance; and (3) a decision mechanism to determine whether the gap is significant and, if so, the corrective action to be taken.

There are, again, a number of conceptual differences in the setting up of control systems for the two major types of planning structures. A principal difference is that evaluation standards for standing plans are derived from a variety of sources, while the control targets for single-use plans are determined largely by the planners. In other words, the effective single-use plan is not complete until a set of benchmarks is established against which implementation will be gauged.

All of the above is well known to professional managers, especially those involved in controlling the large and complex projects typical of NASA. Indeed, by reputation, NASA is considered one of the most competent controllers of project execution, vendor commitment, and overall program progress. It would be extremely presumptive to offer any suggestions for the improvement of NASA controls over the execution of its specified programs.

However, the issue in this study is evaluating and controlling the *impacts* of NASA programs, rather than the progress of the programs themselves. This is a distinctly different concept, and one which has received little attention to date. The traditional approach of NASA to the assessment of its impacts on any segment of society has been essentially a *post-facto* audit or review. The earlier impact studies cited in this book are typical examples of an approach which sets out to determine what happened and to assess the relative benefits and contributions. One of the key themes of this study is that the array of impacts of any NASA program should be assessed a priori and hence considered in its early planning. In other words, to the extent possible, the broadest range of both specific accomplishments and program impacts should be planned in advance, rather than treated as a series of almost random surprises to be discovered in retrospect.

The first phase of impact control should occur during early planning for impacts. As planners identify the kind of impacts to be anticipated at different stages of the program's evolution, and the principal impact groups, they should establish the benchmarks or standards against which impacts are to be measured. Ideally, a series of progressive evaluation targets should be established rather than an all-or-nothing final measure. If, for example, a desired impact is to increase the ease of entry for small firms in a particular industry, a series of control points could be established to measure progress toward that goal. Successive target points might consist of the number of active firms in the industry, industry concentration ratios, geographical dispersion measures, and so forth.

Another useful device in the establishment of control points is to select "early warning" or leading indicators; that is, attempting to identify those tell-tale developments which, if accomplished, would signal an almost certain achievement of the end goal — or, conversely, identify those events or occurrences which are almost sure to precede failure or severe difficulty. By monitoring progress against such early warning indicators, program managers can provide themselves with as long a lead time as possible, and hence greater flexibility in taking corrective action. Experience has shown that the selection of appropriate control points is best accomplished at the time plans are made. At that time, all aspects of the plan and the forecasts are still fresh and alive in the planners' minds. The tracing of probable event chains, probable group responses, and the probability distribution of possible impacts provides the planner with a vivid, dynamic picture of the unfolding scenario. Then is the time to select control points — especially the early warning type.

Once control points have been selected, the control task shifts to one of monitoring and comparing. The selected points provide a focus for the monitoring of an otherwise endless series of events. They make it easier to select certain segments of the economic, technological, social, or political environment for close observation. Even sharper focus is possible if the key impact groups are identified in the setting of control points. At this stage, good use may be made of the spokesmen or sounding boards identified earlier. As experts in what is happening in their particular arena and as valid interpreters of their group's needs and values, each spokesman can be most useful in assessing the width and severity of the gaps between standards expected and performance actually delivered. It also is possible to make use of outside research organizations in determining gaps between anticipated and delivered impacts. Rather than commission a post-mortem assessment, planners would get more utility from an interim study which focused on actual developments that are contrasted against some set of anticipations — planners' and impact groups' anticipations alike.

Finally, as in any control system, there should exist a feedback loop which provides new inputs for improved planning. In the above example, the periodic measuring of the extent to which impact groups' expectations are being met almost automatically provides an updating of the shifts and trends in their needs and values. Whether impact control is exercised through sampling the identified spokesmen, through assigned research projects or through more direct contact with many of the groups, it cannot help but lead to closer familiarity with their changing expectations. Thus successive efforts at impact group analysis tend to improve their validity, increase sensitivity to subtle change, and reduce the time investment required.

A Guide to Impact Management

Impact group analysis, the forecasting of impacts, and a forward-looking impact control system are essential to the task of impact management. A number of specific steps are required to provide NASA with an organized but flexible process for the effective selection and strategic guidance of the programs it undertakes — that is, a process which attempts to plan both specific accomplishments and program impacts for individual projects, and to assess the total mix of impacts of any one of several bundles of programs which together make up NASA's mission at any particular time.

The emphasis is on the long-run effectiveness of NASA as a continuing and evolving organization, rather than on the individual projects in which it engages. Important and spectacular as the specific projects may be, they are viewed here as the tactical components of a complex strategic whole. An organization consisting of a series of essentially one-of-a-kind, non-repetitive events often lacks the recurring patterned missions and activities which gave it permanence and stability. Theorists emphasize the necessity of both stability and innovation within any organization if it is to survive and grow. In fact in most organizations the concern is that too much emphasis on continuity and stability may suppress innovative and creative activities. The concern of NASA should be oppositely directed.

The relationship between individual NASA programs and the agency's total mission is analogous to the relationship between *corporate strategy*, and the *business unit or functional strategy*, in the private sector. Within this framework, the corporation is seen as the enduring, organizational entity which is made up of many temporary businesses. Each of these businesses needs to be managed within a strategic plan devised for its specific missions and functions. The role of the corporate strategy is to coordinate and balance these individual business strategies. While there have been important advances in methodology for designing unit strategy, there has not been comparable progress in conceptualizing the design of corporate strategies.

Not surprisingly, the factors which complicate the task of corporate strategy planning are rather similar to those which confront NASA's total institutional mission. The overall strategy must cope with several business units (programs) of varying size, complexity, and duration. Some of these units may be in the early stage of their life cycle; others may be reaching maturity or declining. The key impact groups for any given unit or program may not be equally important for other units, or may even form a set of competing interests to other programs or units. The corporate strategy must deal with many less tangible goals and objectives than do the individual units, and it must anticipate the diverse impacts of corporate action, which the individual units can frequently ignore. Because of the similarities between NASA's impact planning problem and corporate strategy formulation processes, adaptations and modifications of concepts and approaches recently developed in the still-evolving search for more effective corporate strategy formulation can provide NASA with a basis for action. To simplify terminology, "agency strategy" and "program strategy" are used throughout this discussion to correspond to the analogous concepts of corporate and unit strategies, respectively.

The strategic planning and evaluation process may be visualized as consisting of an interaction between two levels: program strategy and agency strategy. It follows a prescribed, timed pattern (Figure 6.1.) — probably on an annual basis — and consists of the major steps in the discussion below.

A. Impact Group Analysis. The first stage in the proposed process involves a careful analysis of NASA's key impact groups. Once these groups are subdivided into relevant, homogenous categories, their stakes, interests, and expectations of NASA need to be described. The analysis should serve both agency and program strategy formulation. At the upper level (designated A1 in the figure), impact group analysis can be translated into a set of guidelines for NASA's overall strategy and mission. It can be pushed farther, it is hoped, to a statement of "essential" and "desirable" targets and objectives for NASA's total range of operations. At the lower or program level (A2 in the figure), impact group analysis can provide a set of checklist questions against which individual programs can be rated.

B. Agency Strategy Formulation. In the second stage of the process, the agency-level planning unit develops a broad set of overall NASA goals and constraints which are to be used as guidelines for the program-level strategic planners. These guidelines would be derived from agency-level assessment of the ongoing mix of programs, the anticipated political and economic climate, and the aspirations and expectations of NASA's top management team. The guidelines would be rather broad in nature and might suggest specific impact groups for special attention, as well as certain caveats or restrictions.

In addition to the general mission guidelines, agency-level planning might also consider providing program planners with one or more scenarios of its view of the future. Obviously these scenarios should focus on those aspects of the world which would be of interest to program planners. Political forecasts, probable major trends in technologies of selected industries, economic

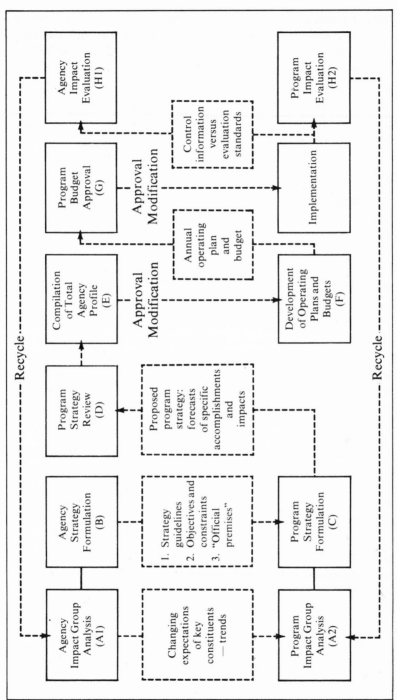

Figure 6.1. Strategic Planning and Evaluation Process

forecasts, and probable issues of future public concern are the kinds of topics which could be covered in a futurist's scenario.

The final item to be provided by the agency-level planners is an outline of questions or procedural steps which program planners could use for program strategy planning and to present the proposed strategy for consideration by top management. It is often advantageous to suggest a format and also to identify the key issues which should be included in the program planner's presentation.

C. Program Strategy Formulation. Armed with guidelines, future outlooks, and procedural directions, the program-level managers develop a program strategy in the third phase of the process. The content of strategic planning in terms of specific accomplishments is somewhat familiar to NASA planners and therefore need not be discussed here. Our emphasis at this stage is to include the forecasting of program impacts and to consider the impacts of programs on the entire range of impact groups, rather than on only the most obvious or direct.

The procedure here is to consider a series of alternative ways of achieving each step in the program and the impacts they will have on the key group. The methods of impact group analysis and impact forecasting have already been described; only the flexibility of the methodologies to be used in dealing with rather uncertain futures needs to be stressed again.

D. Program Strategy Review. The preliminary program strategies formulated in stage C of Figure 6.1 are now presented for agency-level review. The presentations need to include a set of projected impacts (costs and benefits) by group. It might also require the proposing of one or more alternatives which would have significantly different cost-benefit profiles. The presentations would probably be most effective if they combine a brief written report with a longer oral presentation and an opportunity to answer questions and criticisms. The agency-level planners in their review should be concerned with validity, logic consistency, and realism as well as issues of degree of risk, probability of success, and acceptability of the means suggested.

E. Compilation of Total Agency Profile. After each preliminary program strategy has been reviewed in stage D, the agency-level planners prepare a summarized profile which aggregates the impacts of existing programs with the newly proposed ones, or modifications of older ones. Again the impacts need to be classified by group and presented in juxtaposition to their expectations. At the completion of this phase, the agency-level planners present the total agency profile to NASA's top directorate. The presentation will indicate what the total agency impacts are likely to be on key impact groups if NASA were to pursue one set of programs as against another. An important part of this presentation should also be a time-scaled estimate of the latest date by which major commitments must be made if that particular portfolio is adopted. The intent is to provide top management with an indication of the degree of flexibility available with each alternative. It is also a means of injecting as much flexibility as possible into a highly uncertain planning activity.

F. Development of Operating Plans and Budgets. As a result of the review in stages D and E, the agency directorate has selected a portfolio of programs which appears to best meet the needs of the public served by NASA. The individual program strategies submitted by the respective program managers have been approved or modified. The go-ahead has been given for any required major commitments. Now, in stage F, program managers devise operating plans based on the approved strategy, develop budgets, establish personnel needs, and lay out the many operational details involved in implementation. Again, this is an activity highly familiar to NASA management and needs no further elaboration.

G. Program Budget Approval. The annual operating plans and budgets developed for each program in stage F are presented to senior NASA management for final approval. Agency-level planners assist in this process by verifying that the program plan, budget, and other key operational elements are consistent with the previously approved program strategy. In submitting final operating plans, program managers must also identify the benchmarks against which program impacts are to be measured. Since these impacts will be reviewed periodically, as well as on a year-end basis, evaluation targets should also be submitted. After obtaining final approval of program budgets and operating plans, and instituting any modifications suggested by this final review, program managers assume responsibility for the implementation of plans.

H. Impact Evaluation and Control. This final step involves the monitoring or control phase of the process. The specifications of an effective impact control system already have been fully discussed earlier in this chapter. What needs to be added here is that this activity should be carried out at both the program (H2 in Figure 6.1) and agency level (H1 in the figure). The monitoring process will pick up deviations from targets, assumptions which have failed to materialize, and new data previously unknown. All these provide the grist for the next annual review. Agency-level review (step B in the process) results in modified guidelines, while program-level monitoring suggests modifications in unit strategies (step C in the process). The seven-step process is thus recycled and refined.

The proposed approach to impact planning and evaluation has significant organizational and personnel consequences. Implicit in the above recommendation is the need for a high-level, centrally located organizational unit, which should be concerned with the continuity of the organization, its long-term relationship to its environment and its public, and its absorption of a variety of programs into a consistent, stable whole. Restated in the terminology introduced earlier, the crucial concern of this high-level unit should be the formulation of agency strategy.

There are subtle personnel considerations which bear on the staffing of an agency strategy unit. Any organization engaged in the pursuit of largely nonrepetitive activities must be prepared to staff programs and projects with a variety of different (and often unpredictable) skills and experiences. Even within any ongoing program, because of its one-of-a-kind nature, a great flexibility in the available skill mix must be anticipated. Such staffing can be achieved by a combination of measures: by hiring people of broad educational backgrounds and flexible personality, by providing periodic training and retraining or updating, and by rotating assigned personnel. Obviously, NASA relies on contractor personnel and the use of consultants to provide for the changing mix of skills its technical missions demand. It is hoped that it also searches for the breadth and flexibility which should typify its permanent personnel. One of the major tasks of the permanent career professional in NASA is the maintenance of the continuity needed in strategic planning. Career professionals provide not only the historical knowledge of the agency's evolution, but also the extensive contact with and sensitivity to NASA's impact groups.

It is valuable to note here that the impact group analysis described earlier in rather formal, rational, and (possibly) somewhat mechanistic steps is also a highly informal, politically sensitive process. Many decisions are reached through discussion, negotiation, and tradeoff between several competing points of view. To the extent that permanent NASA planning staffs can readily relate to and represent those different perspectives, the impact groups analysis will reflect a consensus of the cost and benefits each group expects. In this way, direct spokesmen within NASA serve to augment and internalize the points of view communicated by the "listening posts" or "barometers" suggested earlier. Such staff representation should not simply reflect the concerns of Congress or the Executive branch, but also the several other groups which contribute to NASA's efforts or are its primary beneficiaries.

A Final Word

A few final words need be said in support of what must sound like a highly complex, slow, and cumbersome process. In the first place: It works! At least it has in many large business corporations which have tried it. Second, great flexibility is built into this system, despite the seemingly mechanical nature of the steps described. It must be remembered that the strategies discussed here would typically involve a five-year forecasting frame. Thus none of the plans, targets, or missions analyzed is expected to evolve as originally projected. Annual reviews would continually adjust current operations and modify the next five-year strategy. Thus planning described here is in the nature of a star by which to steer, rather than the actual target to be hit in five years.

Third, it does *not* take enormous amounts of additional time and energy. Most of the planning, thinking, and exploration proposed here normally take place anyway. It is usually scattered, relatively unorganized, and rarely coordinated. Imposing a formal structure could actually *reduce* rather than *increase* the workload. Fourth, the process may prove painful at first, but experience shows that each subsequent recycling becomes easier, quicker, and more effective. Much learning and adapting need be done, and they can only be done through on-the-job training.

Finally, strategic planning may be an extremely difficult and challenging activity, but there is much persuasive evidence to suggest that it is absolutely essential for success and survival. In NASA's case, it may not be a luxury or a frill, but a basic and vital need.

It would have been illuminating to review the decision-making process involved in a number of past NASA programs and to contrast them with the normative approach recommended in this chapter, and use structured steps of the model presented in the preceding chapter. This exercise, however, would have demanded a significantly different set of efforts and called for rather different bodies of data from those needed for the probes presented in earlier chapters. It would also have required additional time for both information gathering and analysis. It is therefore suggested that such testing of the current recommendations be scheduled for future research and exploration. Testing could proceed along two different, but parallel, paths — a review of earlier program decisions and a monitoring of an ongoing planning effort.

The first effort should reconstruct the decisions made in planning and implementing a recently completed program, and then recast it in terms of the steps proposed here — particularly impact group analysis and impact monitoring. The second, and perhaps more significant, effort would focus on a conscious effort to test the procedures described here in structuring and formulating the basic plans for a new program in its early stages. These procedures could be tested as a parallel, redundant effort, and thus avoid introducing any additional risk to current practice. These tests would, however, provide a secondary structuring of the problems and — more importantly — suggest modifications or elaborations to make application more effective. It is by applying and doing that innovative approaches and processes are sharpened and made to work. Neither theory elaboration nor logic refinement can substitute for test and application.

References

Drucker, Peter F. *Management: Tasks, Responsibilities, Practices.* New York: Harper & Row. 1973.

National Science Foundation. *Official Program Plan for Support of Extramural Research, Fiscal Year 1974.* Washington, D.C.: National Science Foundation. September 1974.

Index